Analysis of Natural Bioactive Compounds in Plant, Food, and Pharmaceutical Products Using Chromatographic Techniques

Analysis of Natural Bioactive Compounds in Plant, Food, and Pharmaceutical Products Using Chromatographic Techniques

Editor

Faiyaz Shakeel

Basel • Beijing • Wuhan • Barcelona • Belgrade • Novi Sad • Cluj • Manchester

Editor
Faiyaz Shakeel
King Saud University
Riyadh, Saudi Arabia

Editorial Office
MDPI
St. Alban-Anlage 66
4052 Basel, Switzerland

This is a reprint of articles from the Special Issue published online in the open access journal *Separations* (ISSN 2297-8739) (available at: https://www.mdpi.com/journal/separations/special_issues/189U457VRA#Editorial).

For citation purposes, cite each article independently as indicated on the article page online and as indicated below:

Lastname, A.A.; Lastname, B.B. Article Title. *Journal Name* **Year**, *Volume Number*, Page Range.

ISBN 978-3-0365-9184-1 (Hbk)
ISBN 978-3-0365-9185-8 (PDF)
doi.org/10.3390/books978-3-0365-9185-8

© 2023 by the authors. Articles in this book are Open Access and distributed under the Creative Commons Attribution (CC BY) license. The book as a whole is distributed by MDPI under the terms and conditions of the Creative Commons Attribution-NonCommercial-NoDerivs (CC BY-NC-ND) license.

Contents

About the Editor vii

Faiyaz Shakeel
Analysis of Natural Bioactive Compounds in Plant, Food, and Pharmaceutical Products Using Chromatographic Techniques
Reprinted from: *Separations* **2023**, *10*, 541, doi:10.3390/separations10100541 1

Ishfaq Mohiuddin, T. Ramesh Kumar, Mohammed Iqbal Zargar, Shahid Ud Din Wani, Wael A. Mahdi, Sultan Alshehri, et al.
GC-MS Analysis, Phytochemical Screening, and Antibacterial Activity of *Cerana indica* Propolis from Kashmir Region
Reprinted from: *Separations* **2022**, *9*, 363, doi:10.3390/separations9110363 5

Abdel-Azeem S. Abdel-Baki, Shawky M. Aboelhadid, Saleh Al-Quraishy, Ahmed O. Hassan, Dimitra Daferera, Atalay Sokmen and Asmaa A. Kamel
Cytotoxic, Scolicidal, and Insecticidal Activities of *Lavandula stoechas* Essential Oil
Reprinted from: *Separations* **2023**, *10*, 100, doi:10.3390/separations10020100 17

Nazrul Haq, Faiyaz Shakeel, Mohammed M. Ghoneim, Syed Mohammed Basheeruddin Asdaq, Prawez Alam, Saleh A. Alanazi and Sultan Alshehri
Greener Stability-Indicating HPLC Approach for the Determination of Curcumin in In-House Developed Nanoemulsion and *Curcuma longa* L. Extract
Reprinted from: *Separations* **2023**, *10*, 98, doi:10.3390/separations10020098 33

Melinda Andrasi, Gyongyi Gyemant, Zsofi Sajtos and Cynthia Nagy
Analysis of Sugars in Honey Samples by Capillary Zone Electrophoresis Using Fluorescence Detection
Reprinted from: *Separations* **2023**, *10*, 150, doi:10.3390/separations10030150 49

Ali Altharawi, Safar M. Alqahtani, Sagar Suman Panda, Majed Alrobaian, Alhumaidi B. Alabbas, Waleed Hassan Almalki, et al.
UPLC-MS/MS Method for Simultaneous Estimation of Neratinib and Naringenin in Rat Plasma: Greenness Assessment and Application to Therapeutic Drug Monitoring
Reprinted from: *Separations* **2023**, *10*, 167, doi:10.3390/separations10030167 59

Nazrul Haq, Faiyaz Shakeel, Mohammed M. Ghoneim, Syed Mohammed Basheeruddin Asdaq, Prawez Alam, Fahad Obaid Aloatibi and Sultan Alshehri
Determination of Pterostilbene in Pharmaceutical Products Using a New HPLC Method and Its Application to Solubility and Stability Samples
Reprinted from: *Separations* **2023**, *10*, 178, doi:10.3390/separations10030178 71

Raheel Suleman, Muawuz Ijaz, Huan Liu, Alma D. Alarcon-Rojo, Zhenyu Wang and Dequan Zhang
Evaluation of Chinese Prickly Ash and Cinnamon to Mitigate Heterocyclic Aromatic Amines in Superheated Steam-Light Wave Roasted Lamb Meat Patties Using QuEChERS Method Coupled with UPLC-MS/MS
Reprinted from: *Separations* **2023**, *10*, 323, doi:10.3390/separations10060323 83

El Mustapha El Adnany, Najat Elhadiri, Ayoub Mourjane, Mourad Ouhammou, Nadia Hidar, Abderrahim Jaouad, et al.
Impact and Optimization of the Conditions of Extraction of Phenolic Compounds and Antioxidant Activity of Olive Leaves (*Moroccan picholine*) Using Response Surface Methodology
Reprinted from: *Separations* **2023**, *10*, 326, doi:10.3390/separations10060326 97

Diana Zasheva, Petko Mladenov, Krasimir Rusanov, Svetlana Simova, Silvina Zapryanova, Lyudmila Simova-Stoilova, et al.
Fractions of Methanol Extracts from the Resurrection Plant *Haberlea rhodopensis* Have Anti-Breast Cancer Effects in Model Cell Systems
Reprinted from: *Separations* **2023**, *10*, 388, doi:10.3390/separations10070388 115

Mayra Beatriz Gómez-Patiño, Juan Pablo Leyva Pérez, Marcia Marisol Alcibar Muñoz, Israel Arzate-Vázquez and Daniel Arrieta-Baez
Rapid and Simultaneous Extraction of Bisabolol and Flavonoids from *Gymnosperma glutinosum* and Their Potential Use as Cosmetic Ingredients
Reprinted from: *Separations* **2023**, *10*, 406, doi:10.3390/separations10070406 131

About the Editor

Faiyaz Shakeel

Prof. Faiyaz Shakeel received his Masters and Ph.D. in Pharmaceutics from Jamia Hamdard (Hamdard University, New Delhi, India). At Jamia Hamdard, he worked on nanoemulsion-based drug delivery systems for some poorly soluble drugs. Then, he became a lecturer at the University of Benghazi (Libya), where he worked on nanoemulsion and self-nanoemulsifying drug delivery systems of some biologically active molecules. In 2011, he was awarded the Young Scientist Award from the Association of Pharmacy Professionals (APP). One of his group's research articles was awarded with the most cited paper award from the European Journal of Pharmaceutics and Biopharmaceutics in March 2012. Currently, he is working as a professor at the Department of Pharmaceutics, College of Pharmacy, King Saud University. At King Saud University, he developed several nanocarrier-based formulations of various drugs. He also developed a double nanoemulsion for a self-nanoemulsifying drug delivery system of 5-fluorouracil. He has very good expertise in the solubilization of drug molecules using cosolvency models. He developed various analytical methods for the determination of various drugs in a variety of sample matrices. His research interests lie in the general area of pharmaceutics and novel drug delivery systems. He is the author of more than 430 journal articles and several book chapters. He also has a US patent. He is Editor/Editorial Board Member of several journals such as Pharmaceutics, Molecules, Separations, Current Drug Delivery, and Pharmaceutical Sciences, among others. He has 11,114 total citations with an H-index of 49 and an i10 index of 237. In 2020, 2021, and 2022, he was named to the Stanford/Elsevier list of the top 2% scientists in the world for both his career (coveted) and a single year.

Editorial

Analysis of Natural Bioactive Compounds in Plant, Food, and Pharmaceutical Products Using Chromatographic Techniques

Faiyaz Shakeel

Department of Pharmaceutics, College of Pharmacy, King Saud University, Riyadh 11451, Saudi Arabia; fsahmad@ksu.edu.sa

1. Introduction

A growing tendency toward the discovery and use of natural bioactive compounds that are the least harmful, have the fewest side effects, and are the most natural for the human body has been noticed during the past few decades [1]. As evidenced by the rise in recent studies on the therapeutic properties of plants, this trend has caused a return of healthcare professionals to nature and plants, but with a modern approach that specifically questions how plants help to heal humans and what their exact effects on the human body are [2]. The medicinal properties of plants are related to their phytochemical makeup, which is a complex matrix with a large number of naturally occurring bioactive molecules that must be distinguished between in order to be identified [3,4]. The separation of natural bioactive chemicals from plants can be accomplished utilizing cutting-edge, high-tech, hyphenated chromatographic approaches, which also provide us with lots of information to be able to identify compounds [5–7]. In order to explain a plant's mechanism of action and therapeutic effect, modern healthcare providers need to be able to link the phytochemical profile of a plant employed in therapy to a biological activity [8]. In addition to plants, natural bioactive substances can be found in a variety of foods and pharmaceuticals [8,9]. As a result, it is crucial to analyze these chemicals in plant, food, and pharmaceutical products [8–10].

In order to identify and analyze natural bioactive compounds in plant, food, and pharmaceutical products, this Special Issue has attempted to compile the latest improvements, advancements, and analytical innovations in chromatographic techniques. Additionally, this Special Issue seeks to enable researchers to link the phytochemical profiles of plants, foods, and pharmaceuticals with proven therapeutic effects, which may later substantiate the health-related claims made for these products. This Special Issue includes 10 research articles focused on the analysis of natural bioactive compounds and phytochemicals in plant, food, and pharmaceutical products using innovative chromatographic techniques.

2. Overview of Published Articles

This Special Issue begins with an article by Mohiuddin et al. [11], who studied the chemical composition and antibacterial effects of *Cerana indica* propolis from the Kashmir region. GC-MS analysis was performed to identify the chemical compounds of Kashmiri propolis, and showed the presence of 68 different phytochemicals in Kashmiri propolis. The ethanolic extract of Kashmiri propolis showed the maximum zone of inhibition against *Staphylococcus aureus*. The findings of this research indicate the presence of various secondary metabolites with distinct pharmacological activities.

Abdel-Baki et al. [12] next assessed the chemical constituents, in vitro cytotoxicity, and scolicidal, acaricidal, and insecticidal activities of *Lavandula steochas* essential oil. The phytoconstituents of *L. steochas* essential oils were detected using spectrometry and gas chromatography techniques. The analyses of *L. steochas* oil showed camphor as being the major compound (58.38%). The oil presented significant cytotoxicity and scolicidal activities. The essential oil also showed 100% adulticidal activity against *R. annulatus* at a

10% concentration, whereas the larvicidal activity was 86.67%. However, the oil showed no insecticidal activity. In addition, *L. steochas* oil demonstrated 100% larvicidal and pupicidal effects. The findings of this work suggest that *L. steochas* essential oil could serve as a potential source of scolicidal, acaricidal, insecticidal, and anticancer agents.

Haq et al. [13], in the next article, developed and validated a greener, stability-indicating, high-performance liquid chromatography (HPLC) approach to determine curcumin (CCM) in an in-house developed nanoemulsion, *Curcuma longa* L. extract, and marketed tablets. The greener HPLC approach was found to be linear, rapid, accurate, precise, and sensitive for measuring CCM. The AGREE approach showed an AGREE score of 0.81 for the proposed HPLC method, which indicated an outstanding greenness profile. The proposed HPLC method successfully determined the CCM in the in-house developed nanoemulsion, *Curcuma longa* extract, and commercial tablets. Furthermore, the greener HPLC method was found to be stability-indicating. The results of this work indicate that CCM can be routinely measured in all studied sample matrices using the greener HPLC method.

The capillary electrophoresis (CE) technique with light-emitting diode-induced fluorescence detection was used for the analysis of sugars in honey samples by Andrasi et al. [14]. The optimized CE technique was applied in the measurement of fructose and glucose via the direct injection of honey samples. The proposed CE technique provides high separation efficiency and sensitivity within a short analysis time. Furthermore, it enables the injection of honeys without sample pretreatment. The findings of this study showed the rapid and sensitive analysis of sugars in honey samples using the CE technique with fluorescence detection.

Altharawi et al. [15] developed and validated an UPLC-MS/MS method for the simultaneous determination of neratinib and naringenin in rat plasma using imatinib as the internal standard (IS). The mass spectra of studied compounds were recorded via the multiple reaction monitoring of the precursor-to-product ion transitions. The proposed UPLC-MS/MS method was found to be linear, selective, precise, accurate, and stable. The proposed method was also found to be eco-friendly for the measurement of neratinib, naringenin, and IS. The analytical results of this work showed that the developed method has implications for its applicability in pharmacokinetic studies in humans to support the therapeutic drug monitoring of combination drugs.

Haq et al. [16] developed a rapid and sensitive HPLC approach for the determination of a natural bioactive compound, pterostilbene (PTT), in commercial capsule dosage form, solubility, and stability samples. The developed HPLC approach was linear, rapid, accurate, precise, and sensitive. The proposed HPLC approach was successfully applied in the measurement of PTT in commercial capsule dosage form, solubility, and stability samples. The results indicated that PTT in commercial products, solubility, and stability samples may be routinely determined using the proposed HPLC approach.

In another article, Suleman et al. [17] used two different spices, Chinese prickly ash and cinnamon, to mitigate the formation of heterocyclic aromatic amines (HAAs) in superheated steam-roasted patties. The findings demonstrated significant differences ($p < 0.05$) in the content of both polar and non-polar HAAs in comparison to the control patties. In cinnamon-roasted and Chinese prickly ash patties, both polar and non-polar HAAs were considerably reduced. The results of this study showed that both spices and superheated steam controlled HAAs to a significant level in lamb meat patties.

El Adnany et al. [18] improved the extraction efficiency of phenolic compounds from olive leaves (*Moroccan picholine*) while minimizing the use of harmful chemicals. Ultrasonic extraction using ethanol was found to be the most effective and environmentally friendly approach. The antioxidant activity of the phenolic compounds of olive leaves was also evaluated. Various parameters, such as the extraction time, solid/solvent ratio, and ethanol concentration (independent variables), were evaluated using a response surface methodology (RSM) based on the Box–Behnken design (BBD) to optimize the extraction conditions. The phenolic compounds of olive leaves were identified using the HPLC-MS technique. Various phenolic compounds, such as hydroxytyrosol, catechin, caffeic acid, vanillin, naringin, oleuropein, quercetin, and kaempferol, were found in high concentrations. The findings of this work showed the efficient extraction of phenolic compounds with great antioxidant activity.

Zasheva et al. [19] studied the effects of *Haberlea rhodopensis* methanol extract fractions on the cell viability and proliferation of two model breast cancer cell lines (MCF7 and MDA-MB231 cells) with different characteristics. In addition to the strong reduction in cell viability, two of the fractions showed a significant influence on the proliferation rate of the hormone receptor expressing MCF7 and the triple-negative MDA-MB231 breast cancer cell lines. The results of this study presented a good background for future studies on the use of myconoside (an active constituent of *Haberlea rhodopensis*) for targeted breast cancer therapy.

In the final article, Gomez-Patino et al. [20] developed a rapid protocol for the extraction and separation of the components of the aerial parts of *Gymnosperma glutinosum*. The chemical compounds of chloroformic and methanolic extracts of *G. glutinosum* were identified using the GC-MS technique. The findings revealed the presence of (−)-α-bisabolol (BIS) as the main component in the chloroformic extract, which was isolated and analyzed via ^{1}H NMR to confirm its presence in *G. glutinosum*. The evaluation of methanolic extracts using the UPLC-MS technique demonstrated the presence of six methoxylated flavones and a group of C20-, C18-hydroxy-fatty acids. The findings of this study concluded that the presence of BIS, an important sesquiterpene with therapeutic skin effects, as well as some antioxidant compounds such as methoxylated flavones and their oils, could play an important role in cosmetology and dermatology formulations.

3. Conclusions and Future Perspectives

In the last few decades, a tremendous amount of research on the analysis of natural bioactive compounds in plants, foods, and pharmaceutical products, using a wide range of chromatography techniques, has been performed. This Special Issue has brought together prominent researchers who have explored the diverse application range of chromatographic techniques in the extraction, separation, identification, and analysis of natural bioactive compounds. This Special Issue provides sufficient information on the analysis of natural bioactive compounds in plant, food, and pharmaceutical products using chromatographic techniques. However, one article reported the pharmacokinetic profile of natural bioactive compounds using the highly sensitive UPLC-MS/MS technique. I believe that further applications of these techniques on the biological samples, pharmacokinetic evaluation, and therapeutic drug monitoring of natural bioactive compounds are still required to explore the clinical applications of these techniques. Furthermore, the correlation of identified bioactive compounds and phytochemicals with their biological activity is required and will add the advantages for future studies.

Acknowledgments: The authors who contributed to this Special Issue are highly appreciated. The reviewers who reviewed the articles of this Special Issue are also thankful for their significant efforts to enhance the quality of this Special Issue.

Conflicts of Interest: The author declares no conflict of interest.

References

1. Ojulari, O.V.; Lee, S.G.; Nam, J.-O. Beneficial effects of bioactive compounds from *Hibiscus sabdariffa* L. on obesity. *Molecules* **2019**, *24*, 210. [CrossRef] [PubMed]
2. Luo, C.; Xu, X.; Wei, X.; Feng, W.; Huang, H.; Liu, H.; Xu, R.; Lin, J.; Han, L.; Zhang, D. Natural medicines for the treatment of fatigue: Bioactive compounds, pharmacology, and mechanisms. *Pharmacol. Res.* **2019**, *148*, 104409. [CrossRef] [PubMed]
3. Fu, Y.; Luo, J.; Qin, J.; Yang, M. Screening techniques for the identification of bioactive compounds in natural products. *J. Pharm. Biomed. Anal.* **2019**, *168*, 189–200. [CrossRef] [PubMed]
4. Ivanovic, M.; Razborsek, M.I.; Kolar, M. Innovative extraction techniques using deep eutectic solvents and analytical methods for the isolation and characterization of natural bioactive compounds from plant material. *Plants* **2020**, *9*, 1428. [CrossRef] [PubMed]

5. Chiriac, E.R.; Chitescu, C.L.; Geana, E.-I.; Gird, C.E.; Socoteanu, R.P.; Boscencu, R. Advanced analytical approaches for the analysis of polyphenols in plants matrices-A review. *Separations* **2021**, *8*, 65. [CrossRef]
6. Ramadan, K.M.A.; El-Beltagi, H.S.; Mohamed, H.I.; Shalaby, T.A.; Galal, A.; Mansour, A.T.; Fotouh, M.M.A.; Bendary, E.S.A. Antioxidant, anti-cancer activity and phytochemicals profiling of *Kigelia pinnaata* fruits. *Separations* **2022**, *9*, 379. [CrossRef]
7. Elwekeel, A.; Hassan, M.H.A.; Almutairi, E.; AlHammad, M.; Alwhbi, F.; Abdel-Bakky, M.S.; Amin, E.; Mohamed, E.I.A. Anti-inflammatory, anti-oxidant, GC-MS profiling and molecular docking analyses of non-polar extracts from five salsola species. *Separations* **2023**, *10*, 72. [CrossRef]
8. Cucu, A.-A.; Baci, G.-M.; Cucu, A.-B.; Dezsi, S.; Lujerdean, C.; Hegedus, I.C.; Bobis, O.; Moise, A.R.; Dezmirean, D.S. Calluna vulgaris as a valuable source of bioactive compounds: Exploring its phytochemical profile, biological activities and apitherapeutic potential. *Plants* **2022**, *11*, 1993. [CrossRef] [PubMed]
9. Pai, S.; Hebbar, A.; Selvaraj, S. A critical look at challenges and future scopes of bioactive compounds and their incorporations in the food, energy, and pharmaceutical sector. *Environ. Sci. Pollut. Res.* **2022**, *129*, 35518–35541. [CrossRef] [PubMed]
10. Hussain, A.; Kausar, T.; Sehar, S.; Sarwar, A.; Ashraf, A.H.; Jamil, M.A.; Noreen, S.; Rafique, A.; Iftikhar, K.; Quddoos, M.Y.; et al. A comprehensive review of functional ingredients, especially bioactive compounds present in pumpkin peel, flesh and seeds, and their health benefits. *Food Chem. Adv.* **2022**, *1*, 100067. [CrossRef]
11. Mohiuddin, I.; Kumar, T.R.; Zargar, M.I.; Wani, S.U.D.; Mahdi, W.A.; Alshehri, S.; Alam, P.; Shakeel, F. GC-MS analysis, phytochemical screening, and antibacterial activity of *Cerana indica* propolis from Kashmir region. *Separations* **2022**, *9*, 363. [CrossRef]
12. Abdel-Baki, A.-A.S.; Aboelhadid, S.M.; Al-Quraishy, S.; Hassan, A.O.; Daferera, D.; Sokmen, A.; Kamel, A.A. Cytotoxic, scolicidal, and insecticidal activities of *Lavandula stoechas* essential oil. *Separations* **2023**, *10*, 100. [CrossRef]
13. Haq, N.; Shakeel, F.; Ghoneim, M.M.; Asdaq, S.M.B.; Alam, P.; Alanazi, S.A.; Alshehri, S. Greener stability-indicating HPLC approach for the determination of curcumin in in-house developed nanoemulsion and *Curcuma longa* L. extract. *Separations* **2023**, *10*, 98. [CrossRef]
14. Andrasi, M.; Gyemant, G.; Sajtos, Z.; Nagy, C. Analysis of sugars in honey samples by capillary zone electrophoresis using fluorescence detection. *Separations* **2023**, *10*, 150. [CrossRef]
15. Altharawi, A.; Alqahtani, S.M.; Panda, S.S.; Alrobaian, M.; Alabbas, A.B.; Almalki, W.H.; Alossaimi, M.A.; Barkat, M.A.; Rub, R.A.; Ullah, S.N.M.N.; et al. UPLC-MS/MS method for simultaneous estimation of neratinib and naringenin in rat plasma: Greenness assessment and application to therapeutic drug monitoring. *Separations* **2023**, *10*, 167. [CrossRef]
16. Haq, N.; Shakeel, F.; Ghoneim, M.M.; Asdaq, S.M.B.; Alam, P.; Alotaibi, F.O.; Alshehri, S. Determination of pterostilbene in pharmaceutical products using a new HPLC method and its application to solubility and stability samples. *Separations* **2023**, *10*, 178. [CrossRef]
17. Suleman, R.; Ijaz, M.; Liu, H.; Alarcon-Rojo, A.D.; Wang, Z.; Zhang, D. Evaluation of Chinese prickly ash and cinnamon to mitigate heterocyclic aromatic amines in superheated steam-light wave roasted lamb meat patties using QuEChERS method coupled with UPLC-MS/MS. *Separations* **2023**, *10*, 323. [CrossRef]
18. El Adnany, E.M.; Elhadiri, N.; Mourjane, A.; Ouhammou, M.; Hidar, N.; Jaouad, A.; Bitar, K.; Mahrouz, M. Impact and optimization of the conditions of extraction of phenolic compounds and antioxidant activity of olive leaves (*Moroccan picholine*) using response surface methodology. *Separations* **2023**, *10*, 326. [CrossRef]
19. Zasheva, D.; Mladenov, P.; Rusanov, K.; Simova, S.; Zapryanova, S.; Simova-Stoilova, L.; Moyankova, D.; Djilianov, D. Fractions of methanol extracts from the Resurrection plant *Haberlea rhodopensis* have anti-breast cancer effects in model cell systems. *Separations* **2023**, *10*, 388. [CrossRef]
20. Gomez-Patino, M.B.; Perez, J.P.L.; Munoz, M.M.A.; Arzate-Vazquez, I.; Arrieta-Baez, D. Rapid and simultaneous extraction of bisabolol and flavonoids from *Gymnosperma glutinosum* and their potential use as cosmetic ingredients. *Separations* **2023**, *10*, 406. [CrossRef]

Disclaimer/Publisher's Note: The statements, opinions and data contained in all publications are solely those of the individual author(s) and contributor(s) and not of MDPI and/or the editor(s). MDPI and/or the editor(s) disclaim responsibility for any injury to people or property resulting from any ideas, methods, instructions or products referred to in the content.

Article

GC-MS Analysis, Phytochemical Screening, and Antibacterial Activity of *Cerana indica* Propolis from Kashmir Region

Ishfaq Mohiuddin [1,*], T. Ramesh Kumar [1], Mohammed Iqbal Zargar [2], Shahid Ud Din Wani [2], Wael A. Mahdi [3], Sultan Alshehri [3], Prawez Alam [4] and Faiyaz Shakeel [3]

1. Department of Zoology, Annamalai University, Annamalainagar 608002, India
2. Department of Pharmaceutical Sciences, School of Applied Science and Technology, University of Kashmir, Srinagar 190006, India
3. Department of Pharmaceutics, College of Pharmacy, King Saud University, Riyadh 11451, Saudi Arabia
4. Department of Pharmacognosy, College of Pharmacy, Prince Sattam Bin Abdulaziz University, Al-Kharj 11942, Saudi Arabia
* Correspondence: lone6449@gmail.com

Abstract: Propolis is a resinous compound produced by honey bees. It contains bioactive molecules that possess a wide range of biological functions. The chemical composition of propolis is affected by various variables, including the vegetation, the season, and the area from which the sample was collected. The aim of this study was to analyze the chemical composition and assess *Cerana indica* propoli's antibacterial efficacy from the Kashmir region. Gas chromatography-mass spectrometry (GC-MS) analysis was used to determine the chemical composition of Kashmiri propolis. A range of bacterial strains was tested for antimicrobial activity using different extracts of propolis by agar well diffusion technique. Propolis was found to be rich in alkaloids, saponins, tannins, and resins. The chemical characterization revealed the presence of 68 distinct phytocompounds using GC-MS, and the most predominant compounds were alpha-D-mannopyranoside, methyl, cyclic 2,3:4,6-bis-ethyl boronate (21.17%), followed by hexadecanoic acid, methyl ester (9.91%), and bacteriochlorophyll-c-stearyl (4.41%). The different extracts of propolis showed specific antibacterial efficacy against multidrug-resistant (MDR) strains viz., *Pseudomonas aeruginosa* (MTCC1688), *Escherichiacoli* (MTCC443), *Klebsiella pneumonia* (MTCC19), *Cutibacterium acnes* (MTCC843), and *Staphylococcus aureus* (MTCC96). The EEKP showed the highest zone of inhibition against *S. aureus* (17.33) at 400 μg mL^{-1}. According to the findings of this study, bee propolis contains a variety of secondary metabolites with various pharmacological activities. Furthermore, because of its broad spectrum of positive pharmacological actions and the fact that it is a promising antibacterial agent, more research on propolis is warranted.

Keywords: *C. indica*; propolis; GC-MS; antibacterial; bioactive compounds; chemical composition

1. Introduction

Propolis (also known as bee glue) derives its name from the Greek word pro, which means "in front of" or "at the entrance to," and polis, which means "community" or "city". This natural substance is a unique mixture of various natural components with distinct properties and aids in the protection of the beehive [1,2]. It is made up of a resinous substance that honeybees collect from the buds and exudates of certain trees and plants, which they then combine with beeswax and the enzyme-rich secretion of bee salivary glands, including β-glycosidase [3–5].

Honeybee colonies frequently employ propolis for hive repairs as a sealant to plug the cracks and restrict the openings, limiting the entry of intruders and maintaining the temperature inside the hive at the ideal level for bees, about 35 °C [6]. In addition, they embalm dead invaders with bee glue to prevent decomposition, protect the bee larvae, store honey, and remove potential sources of microbial infestations [7]. Propolis becomes sticky when above room temperature, but it is hard and brittle at low temperatures [8]. It has a very distinctive and pleasant aroma. Depending on its origin and age, its coloration varies from yellow to red to dark brown. Even the existence of transparent propolis has been reported [9,10]. Propolis has been revealed to have antimicrobial properties [11,12], and it has been used for centuries in traditional medicine and remedies.

The complex chemical content of propolis is constantly changing due to various geographical conditions; generally, raw propolis is composed of plant resin (45–55%), wax (25–35%), essential (5–10%), and aromatic oils (5%), pollen, and other natural components (5%). [13]. Depending on the species of bee, the plant's origin, the habitat, and the storage conditions, thechemical composition of propolis varies [14,15]. So far, over 300 chemical compounds have been isolated and identified from the propolis, including sugars, polyols, hydroxy acids, fatty acids, cardanols, flavan derivatives, triterpenes, prenylated flavanones, anacardic acids, aromatic acids, and their esters, and chalcones, [6,16]. Additionally, due to its complex chemical composition, it has been shown that propolis possesses antimicrobial [17,18], antioxidant [19–21], anti-inflammatory [22,23], and antifungal properties [24]. The bioactivity of propolis has been implicated in its flavonoid, phenolic, diterpenic acid, and aromatic acid contents [25]. Globally, propolis formulations are effective antibacterial agents [26–28]. Propolis is effective against a wide range of bacterial strains, though it is most effective against Gram-positive bacteria and less effective against Gram-negative bacteria [4,29]. The findings of numerous studies have revealed that the antibacterial action is based on obstruction of bacterial movement and the activity of various bacterial enzymes, as well as a weakening of the cytoplasmic membrane [29]. Propolis from several countries, including Spain, Romania, and China, was found to contain fructose, glucose, galactose, stachyose, and sucrose [30]. Propolis contains essential amino acids, such as proline and arginine, which are needed for cell regeneration and can vary depending on the flora and environment around bee colonies [31]. Terpenoids are the volatile components of propolis that are responsible for the odor or scent that propolis emits. It also significantly contributes to propolis essential oil extract's biological effects, such as antibacterial and anti-inflammatory properties [32]. According to Sturm and Ulrih (2020), terpenoids were isolated in propolis for the first time in 2011, and 133 terpenes have been reported in propolis so far. Only 16 alkaloids have been isolated from Brazil and Algeria to date [33]. Hexadecanoic acid, methyl ester, is a fatty acid with antioxidant, hepatoprotective, anticancer, anti-inflammatory, anticoronary, anti-arthritic, antieczemic, antihistaminic, and antiandrogenic properties [34].

Therefore, the current study aims to evaluate the antibacterial activities, physicochemical characterization, phytochemical screening, and gas chromatography–mass spectrometry (GC-MS) analysis of various propolis extracts from the chosen area.

2. Materials and Methods

2.1. Chemicals and Apparatus

All of the chemicals and reagents used were of analytical grade purity. Methanol, ethanol, nutrient agar, Mueller–Hinton agar (MHA), and dimethyl sulfoxide (DMSO) were purchased from Himedia (Mumbai, India). Other chemicals used in this experiment were of the highest quality and commercially available on the market. Streptomycin was also procured from Himedia (Mumbai, India). Throughout the media preparations and measurements, ultrapure water was used.

2.2. Collection of Propolis Sample

Crude propolis samples produced by Ceranaindica bees were obtained from honeybee colonies in Chunt-Waliwar, Ganderbal, Jammu, and Kashmir, India (Figure 1a,b). By scraping the surfaces of the walls, frames, entrances, and coverings, propolis was recovered from the beehive [35]. Prior to analysis, the propolis sample was stored at −20 °C.

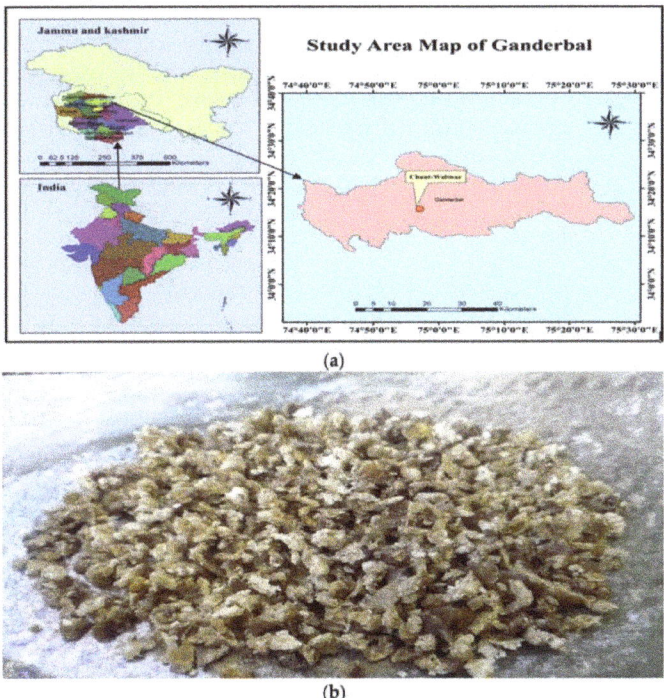

Figure 1. (**a**) The image of map showing the location from where the *C. indica* propolis was collected. (**b**) The image of crude propolis sample produced by *C. indica* honeybees.

2.3. Extraction Procedure

2.3.1. Ethanolic Extract (EEKP)

The 30 g of powdered propolis was extracted using a maceration process with 100 mL of ethanol for at least 3 days. The extract was filtered by using Whatman No.1 filter paper. The extract was dried under pressure using a rotatory evaporator and kept at −20 °C prior to further analysis.

2.3.2. Methanolic Extract (MEKP)

The 30 g of powdered propolis was extracted using a maceration process with 100 mL of methanol for at least 3 days. The extract was filtered using Whatman No. 1 filter paper. The extract was dried under pressure using a rotatory evaporator and kept at −20 °C prior to further analysis.

2.3.3. Aqueous Extract (AqEKP)

About 30 g of Propolis was chopped into pieces and extracted with 100 mL of distilled water. The samples were heated for 5 min on a hot plate with constant agitation at 60 °C before being left at room temperature for a night before being filtered. The resulting mixture was filtered through Whatman filter paper No.1 and concentrated at low temperatures using the freeze-drying method. The extracted samples were kept at −20 °C before the analysis.

2.4. Qualitative Phytochemical Analysis

The different extracts of propolis were subjected to qualitative analysis to identify its constituents, including alkaloids, saponins, tannins, flavonoids, terpenoids, phlobatanins, proteins, and carbohydrates, by following standard protocols [36–39].

2.5. Identification and Quantification of Bioactive Compounds Using GC-MS

GC-MS analysis was performed by using GC-MS/MS-7000D Agilent (Agilent technologies, Santa Clara, CA, USA), equipped with Agilent J&W, GC-MS column HP-5Ms (15 m × 250 mm × 0.25 µm). The carrier gas was helium at a flow rate of 1 mL/min, and the injector temperature was 280 °C in split-less mode. The oven temperature was initially held at 60 °C for 4 min before being increased to 150 °C at a rate of 10 °C/min for 15 min. The following parameters were used to optimize the mass spectra: the source temperature is 280 °C, and the transfer temperature is 150 °C. The solvent delay time was 2 min, and the scan range was 35–500 Da. The temperature was finally raised to 310 °C. The GC's total run time was 40.5 min. By comparing their mass spectra to data from the National Institute of Standards and Technology (NIST) library, the compounds were identified.

2.6. Antibacterial Screening

The agar well diffusion method was used to assess the antibacterial activity of various propolis extracts [40]. After being sterilized at 121 °C for 25 min, all the selected bacterial colonies were first sub-cultured in Nutrient Agar Media 2 (NAM) with a pH of 7. About 4 mL of the NAM was added to test tubes to produce the agar slants, which were then incubated at 37 °C overnight. Tubes containing 15 mL of MHA were inoculated with freshly prepared bacterial inoculums using a sterile loop to ensure uniform distribution of inoculums. The pre-inoculated medium was poured into Petri plates and allowed to solidify before cutting 8 mm wells with a sterile cork borer. The wells were filled with 100 µL of extract at various concentrations (100, 200, and 400 µg mL^{-1}) and an equivalent volume of DMSO, which served as a negative control. After standing for 30 min to allow for pre-diffusion of the extract into the medium, the plates were incubated at 37 °C for 16–20 h. The inhibition zones on the plates were ascertained, and the findings were compared to the positive control containing streptomycin (10 µg mL^{-1}). All assays were carried out in triplicate, and mean values were obtained.

2.7. Bacterial Strains

The five bacterial strains used in this study, viz. *Pseudomonas aeruginosa* (MTCC1688), *Escherichia coli* (MTCC443), *Klebsiella pneumoniae* (MTCC19), *Cutibacterium acnes* (MTCC843), and *Staphylococcus aureus* (MTCC96) were obtained from the Institute of Microbial Technology (IMTECH), Chandigarh, India. As a positive control, streptomycin (10 µg mL^{-1}) was used.

3. Results

3.1. Phytochemical Analysis

The propolis extracts were phytochemically analyzed and found to contain bioactive components, such as terpenoids, flavonoids, alkaloids, phenols, tannins, and saponins. The efficacy of the three solvents in extracting various components of propolis is compared in Table 1.

3.2. GC-MS Analysis

Since EEKP produced the maximum number of phytochemicals when compared to MEKP and AqEKP, GC-MS analysis was performed to determine EEKP's chemical profile. The GC-MS chromatogram of *C. indica* propolis extract shown in Figure 2 shows a total of 68 peaks corresponding to bioactive compounds identified by comparing their mass spectral fragmentation patterns to those of known compounds described in the NIST library. The chemical constituents identified in Table 2 are listed by retention time, peak area (%), and molecular weight. The most common phyto-constituents found in propolis extract are alpha-D-Mannopyranoside, methyl, cyclic 2,3:4,6-bis-ethyl boronate (21.17%), hexadecanoic acid, methyl ester (9.91%), nona-2,3-dienoic acid, ethyl ester (4.75%), bacteriochlorophyll-c-stearyl (4.41%), 10-methyldodecan-4-olide (3.89%), and nickel, cyclopentadienyl-1,2,3-trimethylallyl (3.12%).

Table 1. Phytochemical screening of propolis extracts.

Phytoconstituents	EEKP	MEKP	AqEKP
Alkaloids	++	++	+
Tannins	++	+	+
Saponins	++	++	+
Terpenoids	+	+	-
Flavonoids	++	++	+
Phlobatanins	++	-	-
Anthraquinones	+	+	+
Carbohydrates	++	+	+
Resins	++	+	-
Coumarins	+	+	+
Quinones	++	-	-
Proteins	+	+	-

EEKP: ethanolic extract of Kashmiri propolis; MEKP: methanolic extract of Kashmiri propolis; AqEKP: aqueous extract of Kashmiri propolis; ++: abundant, +: moderate, -: absent.

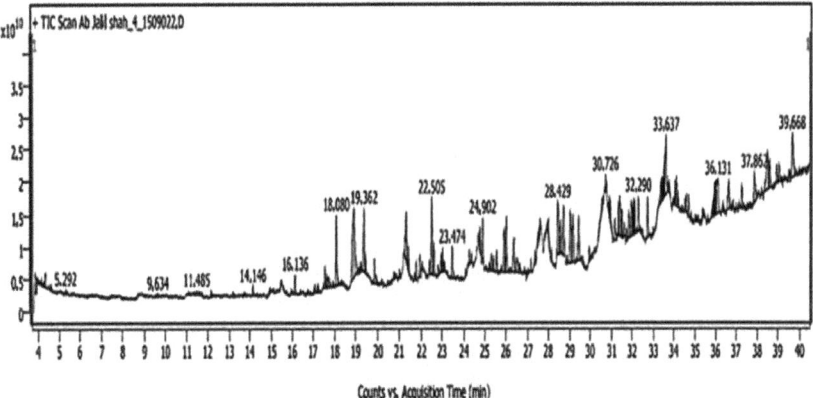

Figure 2. Gas chromatography–mass spectrometry (GC-MS) chromatogram of significant propolis compounds.

Table 2. GC-MS analysis used to identify bioactive compounds in propolis extract (mean ± SD, n = 3.0).

S. N.	Phytocompounds	Molecular Formula	Molecular Weight (g/mol)	Retention Time (min)	CAS Number	Peak Area ± SD (%)
1	2-Azido-2,4,4,6,6-pentamethylheptane	$C_{12}H_{25}N_3$	211.35	3.874	1000293-29-0	1.17 ± 0.01
2	2-Azido-2,4,4,6,6,8,8hepta methylnonane	$C_{16}H_{33}N_3$	267.45	3.974	1000293-29-1	0.44 ± 0.00
3	Methyl 9,10-octadecadienoate	$C_{19}H_{34}O_2$	294.5	4.097	1000336-45-7	0.36 ± 0.00
4	Methyl 8,9-octadecadienoate	$C_{19}H_{34}O_2$	294.5	4.349	1000336-45-0	1.84 ± 0.02
5	Cyclohexane, 1,3-butadienylidene	$C_{10}H_{14}$	134.22	4.620	53864-08-7	0.75 ± 0.01
6	Methyl 10,11-octadecadienoate	$C_{19}H_{34}O_2$	294.5	5.292	1000336-45-3	1.14 ± 0.01
7	1-Propanimine, N-(2-ethylcyclohexyl-, N-oxide	$C_{11}H_{21}NO$	183.29	12.163	1000187-25-8	0.18 ± 0.00
8	Tridecanoic acid, 12-methyl-, methyl ester	$C_{15}H_{30}O_2$	242.4	14.146	5129-58-8	1.55 ± 0.01
9	Decanoic acid, silver(1+) salt	$C_{10}H_{19}AgO_2$	279.12	14.944	13126-67-5	0.98 ± 0.00
10	Methyl 13, 14- octadecadienoate or 13,14-18:2	$C_{19}H_{34}O_2$	294.5	15.961	1000336-46-2	1.42 ± 0.01
11	i-propyl 12-methyl-trideca	$C_{17}H_{34}O_2$	270.5	16.136	1000336-60-4	0.45 ± 0.00
12	4-n-Hexylthiane, s,s- dioxide	$C_{11}H_{22}O_2S$	218.36	16.746	70928-52-8	0.83 ± 0.00
13	Cyclopentyl-methyl-phosphinic acid,2-isopropyl-5-methyl-cyclohexyl ester	$C_{16}H_{31}O_2P$	286.39	17.079	1000194-56-2	0.97 ± 0.00
14	4(axial)-Ethenyl-1,2(equatorial)-dimethyl-trans-decahydroquinol-4-ol,N-oxide	$C_{13}H_{23}NO_2$	225.33	17.734	62299-73-4	0.14 ± 0.00
15	Hexadecanoic acid, methyl ester	$C_{17}H_{34}O_2$	270.5	18.080	112-39-0	9.91 ± 0.12
16	.alpha,-D-Mannopyranoside, methyl, cyclic 2,3:4,6-bis(ethyl boronate)	$C_{11}H_{20}B_2O_6$	269.9	18.900	61553-44-4	21.17 ± 0.20
17	Cis-2-methyl-4-n-butylthiane, s,s-dioxide	$C_{10}H_{20}O_2S$	218.36	19.362	1000215-67-7	2.21 ± 0.02

Table 2. Cont.

S. N.	Phytocompounds	Molecular Formula	Molecular Weight (g/mol)	Retention Time (min)	CAS Number	Peak Area ± SD (%)
18	Borinic acid, diethyl-,1-cyclododecen-1-yl ester	$C_{16}H_{31}BO$	250.2	20.467	61142-73-2	1.11 ± 0.01
19	Trihexadecyl borate	$C_{48}H_{99}BO_3$	735.1	20.751	2665-11-4	0.17 ± 0.00
20	Nona-2,3-dienoic acid, ethyl ester	$C_{11}H_{18}O_2$	182.26	21.323	1000187-19-2	4.75 ± 0.04
21	Methyl 10,11-tetradecadienoate	$C_{15}H_{26}O_2$	238.37	21.710	1000336-31-8	1.95 ± 0.01
22	1-Hexadecanaminium, N,N,N-trimethyl-,octadecanoate	$C_{37}H_{77}NO_2$	568	21.775	124-23-2	0.70 ± 0.00
23	Methyl 8-methyl-decanoate	$C_{12}H_{24}O_2$	200.32	21.943	1000336-49-1	0.21 ± 0.00
24	Tropine N-oxide	$C_8H_{15}NO_2$	157.21	23.019	35722-43-1	1.12 ± 0.01
25	Ethyl trans-4-decenoate	$C_{12}H_{22}O_2$	198.3	23.112	76649-16-6	1.15 ± 0.01
26	1,2-Oxathiane, 6-dodecyl-, 2,2-dioxide	$C_{16}H_{32}O_3S$	304.5	24.395	15224-88-1	0.10 ± 0.00
27	Ethyl hydrogen dimethylamidophosphate, sodium salt	$C_7H_{20}NO_3PSi$	175.1	24.750	1000445-96-4	1.30 ± 0.01
28	9-Borabicyclo[3.3.1]nonane, 9(3-methoxycyclohexyl)oxy-	$C_{15}H_{27}BO_2$	250.19	24.902	1000150-57-9	0.50 ± 0.00
29	1,2:5,6-Di-O-ethylborandiyl-D-glucohexodialdose	$C_{10}H_{16}B_2O_6$	253.9	25.545	74143-57-0	0.23 ± 0.00
30	11-Dodecen-1-ol, 2,4,6-trimethyl-, (R,R,R)-	$C_{15}H_{30}O$	226.4	25.890	27829-54-5	0.11 ± 0.00
31	9-Tetradecenal, (Z)-	$C_{14}H_{26}O$	210.36	26.840	53939-27-8	1.73 ± 0.01
32	10-Methyldodecan-4-olide	$C_{13}H_{24}O_2$	212.33	26.937	1000370-40-6	3.89 ± 0.03
33	Boroxin, tripropyl	$C_9H_{21}B_3O_3$	209.7	27.535	7325-08-8	0.19 ± 0.00
34	Methyl 3,4-tetradecadienoate	$C_{15}H_{26}O_2$	238.37	27.573	1000336-28-1	0.35 ± 0.00
35	Cyclohexanecarboxylic acid, 2-hydroxy-, monoannhydride with 1-butaneboranic acid, cyclic ester, trans-	$C_{11}H_{19}BO_3$	210.1	27.922	24372-06-3	1.16 ± 0.01
36	3,9-Dispiro[5,5]undecane, 3,9-diethyl-, 2,4,8,10-tetraoxa-3,9-dibora-	$C_9H_{18}B_2O_4$	211.9	27.987	58163-67-0	0.82 ± 0.00
37	Methyl myristoleate	$C_{15}H_{28}O_2$	240.38	28.132	56219-06-8	0.38 ± 0.00
38	Bacteriochlorophyll-c-stearyl	$C_{52}H_{72}MgN_4O_4$	841.5	28.429	1000164-49-7	4.41 ± 0.06
39	alpha,-D-Xylofuranose, cyclic 1,2,:3,5-bis(butylboronate)	$C_{13}H_{24}B_2O_5$	282	28.559	52572-01-7	0.16 ± 0.00
40	Z-11-Tetradecenoic acid	$C_{14}H_{26}O_2$	226.35	29.615	1000130-83-3	1.63 ± 0.01
41	Chloroacetic acid, dodecyl ester	$C_{14}H_{27}ClO_2$	262.81	30.061	6316-04-7	0.33 ± 0.00
42	Ethyl 1-(8-amino-1-naphthyl)-1, 2,3-triazole-4-carboxylate	$C_{15}H_{14}N_4O_2$	282.3	30.539	116114-01-3	1.78 ± 0.01
43	Nickel, 2,6,10-dodecatriendi-1,12-diyl-	$C_{12}H_{18}Ni$	220.97	30.901	39330-67-1	0.12 ± 0.00
44	Iron, tricarbonyl-(1,4,5,6,-eta,-4,5-dimethyl-4-hexane-1,6-diyl-1-carboxylic acid, ethyl ester), (endo)-	$C_{14}H_{18}FeO_5$	322.13	30.917	1000149-58-0	0.86 ± 0.00
45	7-Hexadecyn-1-ol	$C_{16}H_{30}O$	238.41	31.179	822-21-9	1.59 ± 0.01
46	Nickel, cyclopentadienyl-1,2,3-trimethylallyl-	$C_{11}H_{16}Ni$	206.94	31.392	77932-64-0	3.12 ± 0.02
47	11,13-Dimethyl-12-tetradecen-1-ol acetate	$C_{18}H_{34}O_2$	282.5	31.608	1000130-81-0	0.73 ± 0.00
48	3-Methylpyrazolobis(diethylboryl) hydroxide	$C_{12}H_{26}B_2N_2O$	236.0	31.715	1000159-71-9	0.37 ± 0.00
49	1-[3-(2,6,6-trimethyl-cyclonex-2-enyl)-4,5-dihydro-3H-pyrazol-4-yl]-ethanone	$C_{14}H_{22}N_2O$	234.34	31.964	1000185-64-0	2.13 ± 0.02
50	Pinane-2,3-diol, 2,3-o-ethaneboronate-	$C_{12}H_{21}BO_2$	208.11	32.077	1000158-69-2	0.21 ± 0.00
51	1H-1,2,3,4-Tetrazole, 5-[(2-methoxyphenoxy)methyl]-	$C_9H_{10}N_4O_2$	206.2	32.102	1000337-41-2	1.24 ± 0.01
52	1,1-Bicyclohexyl, 4-propoxy-4-propyl-	$C_{18}H_{34}O$	266.5	32.290	98321-58-5	1.60 ± 0.01
53	Methyl 10,11-tetradecadienoate	$C_{15}H_{26}O_2$	238.37	32.484	1000336-31-8	0.20 ± 0.00
54	Tris(tert-butyldimethylsilyloxy)arsane	$C_{18}H_{45}AsO_3Si_3$	468.7	33.414	1000366-57-5	1.00 ± 0.01
55	9-Borabicyclo[3.3.1]nonane, 9-[3-(dimethylamino)propyl]-	$C_{13}H_{26}BN$	207.2	33.718	1000160-15-2	2.51 ± 0.03
56	Methyl 12,13-tetradecadienoate	$C_{15}H_{26}O_2$	238.37	33.792	1000336-33-7	0.27 ± 0.00
57	Methyl 8,9-octadecadienoate	$C_{19}H_{34}O_2$	294.5	34.063	1000336-45-0	0.98 ± 0.00
58	Methyl 4,5-tetradecadienoate	$C_{15}H_{26}O_2$	238.37	34.125	1000336-28-7	0.10 ± 0.00
59	.beta,-D-Mannofuranoside, 2,3:5,6-di-ethylboranediyl-cis-nerolidyl	$C_{25}H_{42}B_2O_6$	460.2	34.528	1000155-68-3	0.93 ± 0.00
60	Undec-10-ynoic acid,octyl ester	$C_{19}H_{34}O_2$	294.5	34.599	1000406-16-1	1.88 ± 0.01
61	Methyl 3-cis,9-cis,12-cis-octadecatrienoate	$C_{19}H_{32}O_2$	292.5	35.019	1000336-38-4	0.28 ± 0.00
62	Z,E-2-methyl-3,13-octadecadien-1-ol	$C_{19}H_{36}O$	280.5	35.074	1000131-10-5	0.29 ± 0.00
63	2,2-Dimethyl-6-methylene-1-[3,5-dihydroxy-1-pentenyl]cyclohexan-1-perhydrol	$C_{14}H_{24}O_4$	256.339	35.436	1000212-02-6	0.25 ± 0.00
64	9,12-octadecadienoyl chloride, (z,z)-	$C_{18}H_{31}ClO$	298.9	35.633	7459-33-8	0.34 ± 0.00
65	1-(4-Amino-furazan-3-yl)-5-methoxymethyl-1H-[1,2,3]triazole-4-carboxylic acid hydrazide	$C_7H_{10}N_8O_3$	254.21	36.005	1000300-77-2	0.22 ± 0.00

Table 2. Cont.

S. N.	Phytocompounds	Molecular Formula	Molecular Weight (g/mol)	Retention Time (min)	CAS Number	Peak Area ± SD (%)
66	Iron, tris, (eta,3-2-propenyl)-	$C_9H_{15}Fe$	179.06	37.255	94139-77-2	1.13 ± 0.01
67	E,E,Z-1,3,12-Nonadecatriene-5,14-diol	$C_{19}H_{34}O_2$	294.5	37.342	1000131-11-4	0.26 ± 0.00
68	Phenylboronic acid, 2TMS derivative	$C_{12}H_{23}BOSi_2$	266.29	39.981	7560-51-2	0.30 ± 0.00

3.3. Antibacterial Activity

Table 3 and Figure 3 show the antibacterial properties of different extracts of Kashmiri propolis. EEKP, MEKP, and AqEKP had nearly identical antimicrobial activities. Small differences were observed in the majority of cases (1–3 mm). Only in the cases of *E. coli* and *C. acnes* did MEKP and AqEKP show no activity at 100 µg mL^{-1}.

Table 3. Antimicrobial capacity of various extracts of propolis against tested bacterial strains using agar well diffusion method.

Quantity of Extract (µg/mL)	Zone of Inhibition Diameter (mm)				
	Bacterial Strains				
	S. aureus	P. aeruginosa	K. pneumonae	E. coli	C. acnes
EEKP					
100	12.33 ± 0.33	11.00 ± 0.58	11.33 ± 0.33	09.33 ± 0.33	09.67 ± 0.67
200	14.67 ± 0.67	13.33 ± 0.33	13.67 ± 0.67	11.67 ± 0.33	11.67 ± 0.33
400	17.33 ± 0.67	16.33 ± 0.67	15.67 ± 0.33	13.67 ± 1.20	14.00 ± 0.58
MEKP					
100	11.67 ± 0.88	10.33 ± 0.33	10.00 ± 0.58	-	-
200	14.00 ± 0.58	12.67 ± 0.33	12.00 ± 1.15	11.00 ± 0.58	10.67 ± 0.33
400	16.00 ± 0.58	15.00 ± 0.58	14.33 ± 0.33	13.00 ± 0.58	13.33 ± 0.33
AqEKP					
100	10.67 ± 0.88	07.67 ± 3.84	07.00 ± 3.51	-	-
200	13.00 ± 0.58	11.67 ± 0.33	11.67 ± 0.33	6.33 ± 3.18	10.33 ± 0.33
400	14.67 ± 0.33	13.67 ± 0.67	13.33 ± 0.33	10.67 ± 0.88	12.33 ± 0.33
Streptomycin (10 µg/mL)	19.33 ± 0.88	19.67 ± 0.33	21.00 ± 0.58	18.33 ± 0.33	19.33 ± 0.67

Results in the table are expressed in millimeters (mm) and each value is in triplicate represented as mean ± S.E.M.; - represents no zone of inhibition. EEKP: ethanolic extract of Kashmiri propolis; MEKP: methanolic extract of Kashmiri propolis; AqEKP: Aqueous extract of Kashmiri propolis.

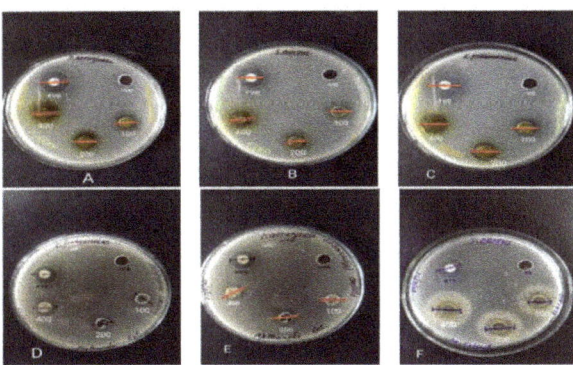

Figure 3. The zone of inhibition (mm) of different extracts of *C. indica* propolis against the selected bacterial strains. (**A**) EEKP vs. *P. aeruginosa*. (**B**) EEKP vs. *S. aureus*. (**C**) EEKP vs. *K. pneumoniae*. (**D**) MEKP vs. *K. pneumoniae*. (**E**) MEKP vs. *P. aeruginosa*. (**F**) AqEKP vs. *S. aureus*. EEKP: ethanolic extract of Kashmiri propolis; MEKP: methanolic extract of Kashmiri propolis; AqEKP: Aqueous extract of Kashmiri propolis.

4. Discussion

The propolis extracts used in the current study showed the existence of a range of bioactive components, including carbohydrates, aldehydes, flavonoids, alkaloids, terpenoids, alcohols, cardiac glycosides, tannins, coumarins, amino acids, phytobatanins, saponins, etc. These phytochemicals could be responsible for the pharmacological properties of the propolis extracts. Propolis is commercialized globally and is considered a significant source of phytochemicals that have pharmacological effects [41]. The presence of 68 compounds from various groups was revealed by the GC-MS analysis, including flavonoids, flavonoid derivatives, terpenes, aromatic acids, and their related esters.

The extracts of propolis have a strong effect against bacteria, such as *Enterococcus* spp., *Escherichia coli*, and *Staphylococcus aureus* [42,43]. In the present studies, the antibacterial activity of various extracts of propolis was tested against an array of bacterial strains. The EEKP was found to possess comparatively stronger antimicrobial activity than MEKP and AqEKP. The highest microbial activity in the EEKP may be due to the highest concentration of flavonoid and phenolic compounds, which may inhibit microorganisms [44]. Another factor is that the majority of naturally occurring secondary metabolites are highly soluble in organic solvents; as a result, the ethanolic extract has the highest microbial activity, followed by MEKP and AqEKP [45]. The propolis extracts had significant antibacterial activity against Staphylococcus aureus but no activity against *E. coli*, according to studies from China and Canada [46,47]. The propolis ethanolic extracts have shown the highest levels of antimicrobial activity against a variety of microorganisms in Brazil [48]. In Vietnam, propolis crude extract demonstrated significant antibacterial activity against S. aureus and inhibited E. coli at lower doses [49]. This is because propolis contains flavones and flavonols that have been isolated from it [50], as well as large levels of terphenyl esters and hydroxybenzoic acid, both of which have antibacterial and antifungal properties [51]. Components of propolis, such as Pinocembrin, exhibit antibacterial action against *Streptococcus* species, whereas p-Coumaric acid, artepillin,3-phenyl-4-di hydrocinnamylocinnamic acid inhibits *H. pylori*, and Apigenin substantially inhibits bacterial glycosyltransferase [52].

5. Conclusions

The present work is the first approach to identify numerous bioactive components by GC-MS analysis and to assess the antibacterial potency of various extracts of *C. indica* propolis from the Kashmir region. The GC-MS analysis of propolis has revealed the presence of 68 bioactive compounds that have a wide range of pharmacological potential, including antibacterial, antifungal, anti-protozoal, antioxidant, hepatoprotective, anti-inflammatory, anticancer, and so on. The presence of these bioactive compounds in propolis is responsible for various therapeutic and pharmacologic properties in traditional medicine. More research is needed to isolate a specific substance that results in a positive result in a biological assay, and appropriate methodologies for in-depth studies should be developed.

Author Contributions: Conceptualization, I.M. and F.S.; methodology, I.M., T.R.K., M.I.Z. and P.A.; software, S.A. and F.S.; validation, S.U.D.W., F.S., S.A. and W.A.M.; formal analysis, S.U.D.W.; investigation, T.R.K. and M.I.Z.; resources, W.A.M. and S.A.; data curation, P.A.; writing—original draft preparation, I.M.; writing—review and editing, F.S., S.A. and W.A.M.; visualization, W.A.M.; supervision, I.M.; project administration, I.M.; funding acquisition, W.A.M. All authors have read and agreed to the published version of the manuscript.

Funding: This research project was supported by Researchers Supporting Project number (RSP2022R516), King Saud University, Riyadh, Saudi Arabia, and APC was supported by the RSP.

Institutional Review Board Statement: Not applicable.

Informed Consent Statement: Not applicable.

Data Availability Statement: Not applicable.

Acknowledgments: Authors are thankful to the Researchers Supporting Project number (RSP2022R516), King Saud University, Riyadh, Saudi Arabia, for supporting this work.

Conflicts of Interest: The authors declare no conflict of interest.

References

1. Park, Y.K.; Alencar, S.M.; Aguiar, C.L. Botanical origin and chemical composition of Brazilian propolis. *J. Agric. Food Chem.* **2002**, *50*, 2502–2506. [CrossRef] [PubMed]
2. Tosic, S.; Stojanovic, G.; Mitic, S.; Pavlovic, A.; Alagic, S. Mineral composition of selected Serbian propolis samples. *J. Apic. Sci.* **2017**, *61*, 5–15. [CrossRef]
3. Al-Ani, I.; Zimmermann, S.; Reichling, J.; Wink, M. Antimicrobial activities of European propolis collected from various geographic origins alone and in combination with antibiotics. *Medicines* **2018**, *5*, 2. [CrossRef]
4. Silva-Carvalho, R.; Baltazar, F.; Almeida-Aguiar, C. Propolis: A complex natural product with a plethora of biological activities that can be explored for drug development. *Evid. Based Complement. Altern. Med.* **2015**, *2015*, 206439. [CrossRef] [PubMed]
5. Papachroni, D.; Graikou, K.; Kosalec, I.; Damianakos, H.; Ingram, V.; Chinou, I. Phytochemical analysis and biological evaluation of selected African propolis samples from Cameroon and Congo. *Nat. Prod. Commun.* **2015**, *10*, 6–70. [CrossRef]
6. Farooqui, T.; Farooqui, A.A. Beneficial effects of propolis on human health and neurological diseases. *Front. Biosci.* **2012**, *4*, 779–793. [CrossRef]
7. Nilesh, K.; Mueen, A.K.; Raman, D.; Ahmed, H. Antioxidant and antimicrobial activity of propolis from Tamilnadu zone. *J. Med. Plant Res.* **2008**, *2*, 361–364.
8. Parolia, A.; Thomas, M.S.; Kundabala, M.; Mohan, M. Propolis and its potential uses in oral health. *Int. J. Med. Med. Sci.* **2010**, *2*, 210–215.
9. Marcucci, M.C.; Rodriguez, J.; Ferreres, F.; Bankova, V.; Groto, R.; Popov, S. Chemical composition of Brazilian propolis from São Paulo state. *Z. Naturforsch. C* **1998**, *53*, 117–119. [CrossRef]
10. Bankova, V.S.; De Castro, S.L.; Marcucci, M.C. Propolis: Recent advances in chemistry and plant origin. *Apidologie* **2000**, *31*, 3–15. [CrossRef]
11. Akbay, E.; Özenirler, Ç.; Çelemli, Ö.G.; Durukan, A.B.; Onur, M.A.; Sorkun, K. Effects of propolis on warfarin efficacy. *Pol. J. Thor. Cardiovas. Surg.* **2017**, *14*, 43–46. [CrossRef]
12. Kakino, M.; Izuta, H.; Tsuruma, K.; Araki, Y.; Shimazawa, M.; Ichihara, K.; Hara, H. Laxative effects and mechanism of action of Brazilian green propolis. *BMC Complement. Altern. Med.* **2012**, *12*, 192. [CrossRef]
13. De Figueiredo, S.M.; Binda, N.S.; Almeida, B.D.M.; Abreu, S.R.L.; De Abreu, J.A.S.; Pastore, G.M.; Sato, H.H.; Toreti, V.C.; Tapia, A.V.; Park, Y.K.; et al. Green propolis: Thirteen constituents of polar extract and total flavonoids evaluated during six years through RP-HPLC. *Curr. Drug Discov. Technol.* **2015**, *12*, 229–239. [CrossRef]
14. Biluca, F.C.; Braghini, F.; Gonzaga, L.V.; Costa, A.C.O.; Fett, R. Physicochemical profiles, minerals and bioactive compounds of stingless bee honey (Meliponinae). *J. Food Compost. Anal.* **2016**, *50*, 61–69. [CrossRef]
15. Lim, D.C.C.; Bakar, M.A.; Majid, M. Nutritional composition of stingless bee honey from different botanical origins. *IOP Conf. Ser. Earth Environ Sci.* **2019**, *269*, 012025. [CrossRef]
16. Popova, M.; Dimitrova, R.; Al-Lawati, H.T.; Tsvetkova, I.; Najdenski, H.; Bankova, V. Omani propolis: Chemical profiling, antibacterial activity and new propolis plant sources. *Chem. Cent. J.* **2013**, *7*, 158. [CrossRef]
17. Velikova, M.; Bankova, V.; Tsvetkova, I.; Kujumgiev, A.; Marcucci, M.C. Antibacterial ent-kaurene from Brazilian propolis of native stingless bees. *Fitoterapia* **2000**, *71*, 693–696. [CrossRef]
18. Barrientos, L.; Herrera, C.L.; Montenegro, G.; Ortega, X.; Veloz, J.; Alvear, M.; Cuevas, A.; Saavedra, N.; Salazar, L.A. Chemical and botanical characterization of Chilean propolis and biological activity on cariogenic bacteria Streptococcus mutans and *Streptococcus sobrinus*. *Braz. J. Microbiol.* **2013**, *44*, 577–585. [CrossRef]
19. Sawaya, A.C.H.F.; Calado, J.C.P.; Santos, L.D.; Marcucci, M.C.; Akatsu, I.P.; Soares, A.E.E.; Abdelnur, P.V.; Cunha, I.B.D.S.; Eberlin, M.N. Composition and antioxidant activity of propolis from three species of Scaptotrigona stingless bees. *J. ApiProd. ApiMed. Sci.* **2009**, *1*, 37–42. [CrossRef]
20. Guimarães, N.S.; Mello, J.C.; Paiva, J.S.; Bueno, P.C.; Berretta, A.A.; Torquato, R.J.; Nantes, I.L.; Rodrigues, T. Baccharisdracunculifolia, the main source of green propolis, exhibits potent antioxidant activity and prevents oxidative mitochondrial damage. *Food Chem. Toxicol.* **2012**, *50*, 1091–1097. [CrossRef]
21. Campos, J.F.; Dos Santos, U.P.; Macorini, L.F.B.; De Melo, A.M.M.F.; Balestieri, J.B.P.; Paredes-Gamero, E.J.; Cardoso, C.A.L.; Souza, K.D.P.; Dos Santos, E.L. Antimicrobial, antioxidant and cytotoxic activities of propolis from Meliponaorbignyi (Hymenoptera, Apidae). *Food Chem. Toxicol.* **2014**, *65*, 374–380. [CrossRef] [PubMed]
22. Barbarić, M.; Mišković, K.; Bojić, M.; Lončar, M.B.; Smolčić-Bubalo, A.; Debeljak, Ž.; Medić-Šarić, M. Chemical composition of the ethanolicpropolis extracts and its effect on HeLa cells. *J. Ethnopharmacol.* **2011**, *135*, 772–778. [CrossRef]
23. Cavendish, R.L.; De Souza Santos, J.; Neto, R.B.; Paixão, A.O.; Oliveira, J.V.; De Araujo, E.D.; Silva, A.E.B.E.; Thomazzi, S.M.; Cardoso, J.C.; Gomes, M.Z. Antinociceptive and anti-inflammatory effects of Brazilian red propolis extract and formononetin in rodents. *J. Ethnopharmacol.* **2015**, *173*, 127–133. [CrossRef]
24. Viuda-Martos, M.; Ruiz-Navajas, Y.; Fernández-López, J.; Pérez-Álvarez, J.A. Functional properties of honey, propolis, and royal jelly. *J. Food Sci.* **2008**, *73*, 117–124. [CrossRef] [PubMed]

25. Coneac, G.; Gafițanu, E.; Hădărugă, D.I.; Hădărugă, N.G.; Pînzaru, I.A.; Bandur, G.; Ursica, L.; Paunescu, V.; Gruia, A. Flavonoid contents of propolis from the west side of Romania and correlation with the antioxidant activity. *Chem. Bull. Politeh. Univ.* **2008**, *53*, 56–60.
26. Da SilveiraRegueira-Neto, M.; Tintino, S.R.; Da Silva, A.R.P.; Costa, M.D.S.; Oliveira-Tintino, C.D.D.M.; Boligon, A.A.; Menezes, I.R.A.; Balbino, V.D.Q.; Coitinho, D.D.M. Comparative analysis of the antibacterial activity and HPLC phytochemical screening of the Brazilian red Propolis and the resin of Dalbergiaecastaphyllum. *Chem. Biodivers.* **2019**, *16*, 1900344.
27. Bogdanov, S. Propolis: Biological properties and medical applications. In *The Propolis Book*; Chapter 2; Publisher Bee Product Science: Bern, Switzerland, 2017; pp. 1–41. Available online: https://www.researchgate.net/publication/304012147_Propolis_biological_properties_and_medical_applications (accessed on 8 September 2022).
28. Isidorov, V.A.; Szczepaniak, L.; Bakier, S. Rapid GC/MS determination of botanical precursors of Eurasian propolis. *Food Chem.* **2014**, *142*, 101–106. [CrossRef]
29. Mirzoeva, O.K.; Grishanin, R.N.; Calder, P.C. Antimicrobial action of propolis and some of its components: The effects on growth, membrane potential and motility of bacteria. *Microbiol. Res.* **1997**, *152*, 239–246. [CrossRef]
30. Qian, W.L.; Khan, Z.; Watson, D.G.; Fearnley, J. Analysis of sugars in bee pollen and propolis by ligand exchange chromatography in combination with pulsed amperometric detection and mass spectrometry. *J. Food Compost. Anal.* **2008**, *21*, 78–83. [CrossRef]
31. Mulyati, A.H.; Sulaeman, A.; Marliyati, S.A.; Rafi, M.; Fikri, A.M. Phytochemical analysis and antioxidant activities of ethanol extract of stingless bee propolis from Indonesia. *AIP Conf. Proc.* **2020**, *2243*, 30014.
32. Bankova, V.; Popova, M.; Trusheva, B. Propolis volatile compounds: Chemical diversity and biological activity: A review. *Chem. Cent. J.* **2014**, *8*, 28. [CrossRef]
33. Šturm, L.; Ulrih, N.P. Advances in the propolis chemical composition between 2013 and 2018: A review. *eFood* **2020**, *1*, 24–37. [CrossRef]
34. Krishnamoorthy, K.; Subramaniam, P. Phytochemical profiling of leaf, stem, and tuber parts of *Solenaamplexicaulis* (Lam.) Gandhi using GC-MS. *Int. Sch. Res. Not.* **2014**, *2014*, 567409.
35. Marletto, F. *Propolis Characteristics in Relation to the Floral Origin and the Bees' Utilization of It*; Apicoltore Moderno: Turin, Italy, 1983; pp. 187–191.
36. Harborne, A.J. *Phytochemical Methods a Guide to Modern Techniques of Plant Analysis*, 3rd ed.; Chapman & Hall: London, UK, 1998; Volume XIV, 302p.
37. Misra, C.S.; Pratyush, K.; Sagadevan, L.D.M.; James, J.; Veettil, A.K.T.; Thankamani, V. A comparative study on phytochemical screening and antibacterial activity of roots of Alstoniascholaris with the roots, leaves and stem bark. *Int. J. Pharm. Pharm. Res.* **2011**, *1*, 77–82.
38. Vijayalakshmi, R.; Ravindhran, R. Preliminary comparative phytochemical screening of root extracts of *Diospyrusferrea* (Wild.) Bakh and *Aervalanata* (L.) Juss. Ex Schultes. *Asian J. Plant Sci. Res.* **2011**, *2*, 581–587.
39. Gul, R.; Jan, S.U.; Faridullah, S.; Sherani, S.; Jahan, N. Preliminary phytochemical screening, quantitative analysis of alkaloids, and antioxidant activity of crude plant extracts from Ephedra intermedia indigenous to Balochistan. *Sci. World J.* **2017**, *2017*, 5873648. [CrossRef]
40. Dhilna, C.R.; Gopinath, S.M.; Sajith, A.M.; Savitha, B.; Shruthi, S.D.; Joy, M.N. Some imidazolylbenzamides as potent antibacterial agents. *AIP Conf. Proc.* **2020**, *2280*, 030004.
41. Miyataka, H.; Nishiki, M.; Matsumoto, H.; Fujimoto, T.; Matsuka, M.; Satoh, T. Evaluation of propolis. I. Evaluation of Brazilian and Chinese propolis by enzymatic and physico-chemical methods. *Biol. Pharm. Bull.* **1997**, *20*, 496–501. [CrossRef]
42. Noori, A.L.; Al-Ghamdi, A.; Ansari, M.J.; Al-Attal, Y.; Salom, K. Synergistic effects of honey and propolis toward drug multi-resistant Staphylococcus aureus, *Escherichia coli* and *Candida albicans* isolates in single and polymicrobial cultures. *Int. J. Med. Sci.* **2012**, *9*, 793.
43. Kasiotis, K.M.; Anastasiadou, P.; Papadopoulos, A.; Machera, K. Revisiting Greek propolis: Chromatographic analysis and antioxidant activity study. *PLoS ONE* **2017**, *12*, e0170077. [CrossRef]
44. Mercan, N.; Kivrak, I.; Duru, M.E.; Katircioglu, H.; Gulcan, S.; Malci, S.; Acar, G.; Salih, B. Chemical composition effects onto antimicrobial and antioxidant activities of propolis collected from different regions of Turkey. *Ann. Microbiol.* **2006**, *56*, 373–378. [CrossRef]
45. Gupta, A.; Naraniwal, M.; Kothari, V. Modern extraction methods for preparation of bioactive plant extracts. *J. Appl. Nat. Sci.* **2012**, *1*, 8–26.
46. Ding, Q.; Sheikh, A.R.; Gu, X.; Li, J.; Xia, K.; Sun, N.; Wu, R.A.; Luo, L.; Zhang, Y.; Ma, H. Chinese Propolis: Ultrasound-assisted enhanced ethanolic extraction, volatile components analysis, antioxidant and antibacterial activity comparison. *Food Sci. Nutr.* **2021**, *9*, 313–330. [CrossRef] [PubMed]
47. Allan, R. Antibacterial activity of propolis and honey against *Staphylococcus aureus* and *Escherichia coli*. *Afr. J. Microbiol. Res.* **2010**, *4*, 1872–1878.
48. Silva, R.P.D.; Machado, B.A.S.; Barreto, G.D.A.; Costa, S.S.; Andrade, L.N.; Amaral, R.G.; Carvalho, A.A.; Padilha, F.F.; Barbosa, D.D.V.; Umsza-Guez, M.A. Antioxidant, antimicrobial, antiparasitic, and cytotoxic properties of various Brazilian propolis extracts. *PLoS ONE* **2017**, *12*, e0172585.

49. Georgieva, K.; Popova, M.; Dimitrova, L.; Trusheva, B.; Thanh, L.N.; Phuong, D.T.L.; Lien, N.T.P.; Najdemski, H.; Bankova, V. Phytochemical analysis of Vietnamese propolis produced by the stingless bee Lisotrigonacacciae. *PLoS ONE* **2019**, *14*, e0216074. [CrossRef]
50. Jug, M.; Karas, O.; Kosalec, I. The influence of extraction parameters on antimicrobial activity of propolis extracts. *Nat. Prod. Commun.* **2017**, *12*, 47–50. [CrossRef]
51. Popova, M.; Trusheva, B.; Antonova, D.; Cutajar, S.; Mifsud, D.; Farrugia, C.; Tsvetkova, I.; Najdenski, H.; Bankova, V. The specific chemical profile of Mediterranean propolis from Malta. *Food Chem.* **2011**, *126*, 1431–1435. [CrossRef]
52. Martinotti, S.; Ranzato, E. Propolis: A new frontier for wound healing? *Burns Trauma* **2015**, *3*, 9. [CrossRef]

Article

Cytotoxic, Scolicidal, and Insecticidal Activities of *Lavandula stoechas* Essential Oil

Abdel-Azeem S. Abdel-Baki [1], Shawky M. Aboelhadid [2,*], Saleh Al-Quraishy [3], Ahmed O. Hassan [4], Dimitra Daferera [5], Atalay Sokmen [6] and Asmaa A. Kamel [2]

[1] Zoology Department, Faculty of Science, Beni-Suef University, Beni-Suef 62511, Egypt
[2] Parasitology Department, Faculty of Veterinary Medicine, Beni-Suef University, Beni-Suef 62511, Egypt
[3] Zoology Department, College of Science, King Saud University, Riyadh 11564, Saudi Arabia
[4] Department of Medicine, Washington University School of Medicine, St. Louis, MO 63110, USA
[5] Laboratory of General Chemistry, Agricultural University of Athens, Iera Odos 75, 118 55 Athens, Greece
[6] Department of Plant Production and Technologies, Faculty of Agriculture and Natural Sciences, Konya Food and Agriculture University, Konya 42080, Turkey
* Correspondence: shawky.abohadid@vet.bsu.edu.eg

Abstract: Essential oils (EOs) have recently attracted more interest due to their insecticidal activities, low harmfulness, and rapid degradation in the environment. Therefore, *Lavandula stoechas* (*L. stoechas*) essential oil was assessed for its chemical constituents, in vitro cytotoxicity, and scolicidal, acaricidal, and insecticidal activities. Using spectrometry and gas chromatography, the components of *L. stoechas* EOs were detected. Additionally, different oil concentrations were tested for their anticancer activities when applied to human embryonic kidney cells (HEK-293 cells) and the human breast cancer cell line MCF-7. The oil's scolicidal activity against protoscolices of hydatid cysts was evaluated at various concentrations and exposure times. The oil's adulticidal, larvicidal, and repelling effects on *R. annulatus* ticks were also investigated at various concentrations, ranging from 0.625 to 10%. Likewise, the larvicidal and pupicidal activities of *L. stoechas* against *Musca domestica* were estimated at different concentrations. The analyses of *L. stoechas* oil identified camphor as the predominant compound (58.38%). *L. stoechas* oil showed significant cytotoxicity against cancer cells. All of the tested oil concentrations demonstrated significant scolicidal activities against the protoscoleces of hydatid cysts. *L. stoechas* EO (essential oil) showed 100% adulticidal activity against *R. annulatus* at a 10% concentration with an LC_{50} of 2.34%, whereas the larvicidal activity was 86.67% and the LC_{50} was 9.11%. On the other hand, the oil showed no repellent activity against this tick's larva. Furthermore, *L. stoechas* EO achieved 100% larvicidal and pupicidal effects against *M. domestica* at a 10% concentration with LC_{50} values of 1.79% and 1.51%, respectively. In conclusion, the current work suggests that *L. steochas* EO could serve as a potential source of scolicidal, acaricidal, insecticidal, and anticancer agents.

Keywords: lavender oil; cytotoxic; scolicidal; *Musca domestica*; acaricide

1. Introduction

In developing countries, parasitic diseases, associated with both ectoparasites and endoparasites, represent a severe hazard to both human and animal health [1]. Cystic echinococcosis (CE) is caused by the larval stage (hydatid cyst) of *Echinococcus granulosus*, which can develop in the liver, heart, lungs, brain, spleen, bone, and kidneys of the host, and can be fatal [2,3]. In several endemic areas, the incidence rate of CE may vary from 1 to 200 per 100,000 persons annually [4]. Currently, a range of chemical scolicidal compounds, such as benzimidazole derivatives, are used to deactivate hydatid cyst protoscolices [5,6]. However, these chemicals cause a variety of negative side effects, including impairments in liver function, leucopenia, and abdominal pain [6,7]. A perfect scolicidal agent is one that remains stable after being diluted with cyst fluid, eliminates cyst protoscolices, is non-toxic, causes no harm to the tissue of the host, is inexpensive, and is easily accessible [3,7–9].

The use of scolicidal compounds is essential to the therapeutic treatment of hydatid cysts and helps prevent the spread of the protoscoleces through surgery [2]. Due to the adverse effects often associated with the use of protoscolicidal agents during hydatid cyst surgery, more emphasis is now placed on the toxicity of these agents and the search for safe alternatives [10]. Significant amounts of research have recently been directed towards examining herbal extracts as a source of novel, powerful, and non-toxic anti-scolicidal compounds [11]. Numerous plant extracts and their essential oils, including *Mentha pulegium*, *Curcuma longa*, *Allium sativum*, *Nigella sativa*, *Zataria multiflora*, *Salvadora persica*, *Origanum minutiflorum*, and *Zingiber officinale* have been shown to carry significant scolicidal effects [2,8,11–13].

Globally, ectoparasites pose a significant risk to both the economy and animal health [14]. Ectoparasitic infestation has been linked to a variety of health issues, including anemia, weight loss, abscesses, and tissue damage. It can also serve as a vector for several deadly diseases of great concern to livestock [14,15]. Among the varieties of ectoparasites, ticks are considered a main threat due to the severe irritation, anemia, paralysis, and toxicosis they can cause, as well as the fact that they can transmit diseases such as anaplasmosis, theileriosis, and babesiosis [16–19]. Similarly, one of the most globally prevalent arthropods with medicinal and veterinary significance is the house fly (*Musca domestica* L.) [20,21]. This species is known to harbor more than 100 types of microorganisms, including bacteria, viruses, parasites, worms, and protozoa, which can lead to serious and potentially fatal diseases in both humans and domestic animals [22,23].

Chemically derived drugs are the primary approach used globally to manage ectoparasites and endoparasites affecting different kinds of animals. This has resulted in several serious problems, including the development of resistance [24,25], toxic damage to non-specified organisms, and environmental pollution [26]. Consequently, new eco-friendly alternatives are being introduced into strategic parasite-monitoring programs [27]. The usage of essential oils and plant extracts as insect control agents has become the subject of intensive investigation in a number of countries because of the efficiency of their insecticidal and acaricidal effects, which have negligible environmental impacts [28,29]. These oils comprise combinations of chemical substances that are toxic to insects, and toxicity operates via a number of mechanisms including enzyme inhibition and protein denaturation [30]. It is known that several plants in the Mediterranean region possess insecticidal and acaricidal properties [31]. The genus "*Lavandula*" (Lamiaceae) is a wild plant found in the Mediterranean basin that comprises over 34 species and is well-known for having insecticidal effects against different species [32,33]. Along with being employed in conventional treatment, different species of genus "*Lavandula*" are also utilized in the pharmaceutical and cosmetic sectors [34,35]. One of the most widely studied and used lavender species in the world is *Lavandula stoechas* (L. stoechas) [36]. Some studies have focused on the antibacterial [37,38], antifungal [39,40], and antioxidative [40,41] characteristics of *L. stoechas*. Correspondingly, the objectives of the current study were to determine the following: (i) the total chemical constituents; (ii) the in vitro cytotoxic activities of *L. steochas*; (iii) the in vitro scolicidal activity of *L. steochas* against the protoscoleces of hydatid cysts; (iv) the in vitro adulticidal, larvicidal, and repellent activities of *L. steochas* against *R. annulatus* ticks; and (v) the in vitro larvicidal and pupicidal activities of *L. steochas* against *M. domestica*.

2. Materials and Methods

2.1. Plant Material

In August 2017, the aerial components of *Lavandula stoechas* subsp. Stoechas were gathered from the city of Elmal-Antalya, and identification was performed by the plant taxonomist Dr. Aşkn Akpulat from the Faculty of Education, Cumhuriyet University, Sivas, Turkey. A voucher specimen was deposited at the Herbarium of the Department of Biology, Cumhuriyet University (CUFH). The leaves of the gathered plants were removed from the stems and flowers, dried in the shade, and then crushed until they could pass through a 2 mm mesh.

2.2. Essential Oil Extraction

Dried and finely crushed leaves (100 g) were hydrodistilled for 3 h in a Clevenger-type distillation apparatus with 2 L of double-distilled water [41,42]. The produced EOs were filtered, dried over anhydrous sodium sulfate, and kept at 4 °C until use.

GC–MS analysis of Lavandula stoechas essential oil was performed using a Trace Ultra gas chromatograph, (GC) coupled with a DSQ II mass spectrometer (MS; Thermo Scientific). The compounds were separated on a TR-5MS (30 m × 0.25 mm × 0.25 µm) capillary column (Thermo Scientific), operating on a temperature program of 60 to 250 °C with an elevation speed of 3 °C/min and a helium flow rate of 1 mL/min. The injector and MS transfer line temperatures were set at 220 and 250 °C, respectively. The samples were prepared via the dilution of 1 mg of EO in 1 mL of acetone. In total, 1 µL of the diluted sample was injected manually in the splitless mode. The MS was operated in the EI mode at 70 eV. The ion source temperature was 240 °C, and the mass spectra were acquired in the scan mode based on a mass range of 35–400. We tentatively identified the compounds based on comparisons of their relative retention indexes and mass spectra with corresponding data found in the literature and different databases [42]. A series of n-alkanes (C8–C24) was used in the determination of the relative retention index (RRI). Relative percentages of the compounds were obtained electronically from area percentage data.

2.3. Cytotoxic Activity of L. stoechas

Cell Culture

Human embryonic kidney cells (HEK-293 cells) and the human breast cancer cell line MCF-7 were cultured in a DMEM culture medium accompanied with 10% fetal bovine serum (FBS), 0.2% sodium bicarbonate, and an antibiotic/antimycotic solution. The cells were grown in a CO_2 incubator (5% CO_2–95% atmosphere) at ±37 °C and a high humidity [43]. The trypan blue dye exclusion assay was used to determine the vitalities of all cell lines, and batches of cells with over 98% cell viability were utilized.

2.4. Cytotoxicity Assessment by MTT Assay (3-(4,5-dimethylthiazol-2-yl)-2,5-diphenyltetrazolium Bromide Tetrazolium)

The cytotoxicity assessment was performed according to the protocol set out by Siddiqui et al. [43]. In brief, the cells were plated in 96-well culture plates and adhered for 24 h in a CO_2 incubator at ±37 °C. Then, the cells were exposed to *L. steochas* oil in several concentrations (0.0156–1%) for 24 h. Following exposure, 10 mL of MTT (5 mg/mL of stock) was added to each well, and the plates were then incubated for a further 4 h in a CO_2 incubator. After the supernatant was discarded, 200 mL of DMSO was added to each well and thoroughly mixed. The plates were read at a wavelength of 550 nm. The cytotoxic activity of *L. steochas* oil against the cancer cell line was inferred from the estimated LC_{50} value.

2.5. Scolicidal Activity

Collection of Protoscoleces from Hydatid Cysts and the Viability Test

Using sheep livers that were naturally infected with hydatid cysts, the careful excision of the hydatid was performed. The hydatid fluid samples were aspirated with protoscoleces and kept in glass cylinders for 20 min to allow the protoscoleces to settle. The supernatant fluid was then discarded and the reaming protoscoleces were washed three times in saline solution. An eosin stain at 0.1% was used to confirm the viability of the protoscoleces. Staining was performed for five minutes and the protoscoleces that did not take on the dye were considered alive, while those that were stained were considered dead [44]. Protoscoleces with a viability of 95% were selected for further study.

2.6. Determination of In Vitro Scolicidal Activity

To assess the in vitro activity, the protoscoleces treated with three concentrations of essential oil (0.025, 0.05, and 0.1%) were evaluated. A total of 2 mL of each concentration was placed into three tubes to which ~5×10^3 protoscolices were added, which were gently mixed. Then, the three tubes of each concentration (nine in total) were incubated at ± 37 °C for 1, 3, and 5 min, respectively. After incubation, the supernatant was carefully removed to avoid unsettling the remaining protoscoleces. Then, the settled protoscoleces were gently mixed with 1 mL of 0.1% eosin. Stained protoscoleces were then examined under a microscope five minutes later to determine their vitality. A group of at least 5×10^3 protoscoleces in 2 mL of distilled water with no oil treatment was used as the control.

2.7. Acaricidal Activity of L. steochas against Larvae and Adult of R. annulatus

In various villages in the Beni-Suef province of central Egypt, and further south towards Cairo (29°04′N, 31°05′E), adult engorged females of R. annulatus were collected from naturally infected cattle. The collected ticks were taken to the Parasitology Lab at the Faculty of Veterinary Medicine, Beni-Suef University. The ticks were identified according to the process of Estrada-Pena et al. [45]. A portion of the collected ticks were employed in the adult immersion test, while the remaining were incubated at 27 ± 1.5 °C and 70–80% relative humidity (RH) to produce eggs, which were then allowed to develop into larvae for use in subsequent bioassays.

2.8. Adult Immersion Test of R. annulatus

L. stoechas essential oil was evaluated for adulticidal activities against adult female R. annulatus ticks taken from naturally infected cattle. This assay was performed following the method of Drummond et al. [46]. This test was performed over five replicates (ten ticks/replicate) for each concentration. The ticks of the control group were treated with ethanol 70% and deltamethrin 5%. Briefly, the ticks were immersed in 10 mL of each concentration in a Petri dish with a diameter of 7 cm at room temperature with occasional gentle agitation. After 2 min, the solution was discarded, and the female ticks were removed and gently dried on a paper towel. For ovipositioning, the treated ticks were maintained in a BOD incubator at a temperature of 27 ± 2 °C and a relative humidity of $80 \pm 10\%$. On day 14 post-application (PA), the eggs deposited by the treated ticks were collected and weighed. After 14 days, the number of dead adult ticks was determined, and the egg production index (EPI) was calculated for the ticks that were still alive [47,48]. Egg production index (EPI) = weight of egg mass/initial weigh of engorged female $\times$ 100.

2.9. Larvicidal Activity against R. annulatus

The larvicidal activity of L. stoechas essential oil was assessed via application against the larvae of R. annulatus ticks using the larval packet technique (LPT) with the modifications suggested by Matos et al. [49]. In brief, about one hundred ten-day-old larvae were placed on the center of 7×7 cm filter paper, which was impregnated with 100 µL of each concentration and then folded into a pocket shape. After 24 h, the packets were checked to assess the mortality rates—motionless larvae were considered dead. The filter paper of the control groups was impregnated with ethanol (70%) and deltamethrin 5%. This test was conducted three times for each concentration.

2.10. Repellent Activity against R. annulatus

This bioassay was based on the vertical migration behavior of tick larvae and was modeled after that elucidated by Wanzala et al. [50] with some modifications. In this assay, we used a device consisting of two aluminum rods (0.7×15 cm) and filter paper (7×7 cm) impregnated with 200 µL (covering approximately 28 cm^2) of the different concentrations. The treated filter paper was clipped to one rod, while on the other rode, filter paper impregnated with 70% ethanol was clipped, and this acted as a negative control.

Nearly 30 ten-day-old *R. annulatus* larvae were placed at the base of each rod; the rods were observed after 15 min and after 1 h, and then followed up 4 h post-application. Larvae that were found on the tops of the impregnated filter paper were not considered repelled, while those at the base of the impregnated filter paper (the uncovered part of the rod) were considered repelled. This test was performed five times for each concentration.

$$\text{The repellence (\%)} = \frac{NC - NT}{NC} \times 100$$

NC = number of larvae on the negative control; NT = number of larvae on the treated paper

2.11. Insecticidal Activity against Musca domestica

Rearing of Housefly Colony

Adult house flies were collected from a farm in Beni-Suef province, Egypt. The collected house flies were taken to the Parasitology Lab at the Faculty of Veterinary Medicine, Beni-Suef University. The flies were kept in plastic jars (35 × 15 cm) at $28 \pm 2\ °C$ and 60–70% relative humidity (RH), covered with muslin cloth. A cotton swab soaked in milk (10% w/v) was introduced as food to the adult flies, and this also served as a substratum for oviposition. For hatching and larval development, the eggs were transferred to a different set of jars containing animal feed or cotton swabs soaked in milk. Similarly, pupae were collected and maintained in a separate container until they emerged as adults. Larvae and pupae were used in the bioassays, as recommended by Jesikha [51] and Abdel-Baki et al. [52].

2.12. Larvicidal Bioassay against Musca domestica

The residual film method, as set out by Busvine [53] and modified by Palacios et al. [54], was used to evaluate the larvicidal activity of *L. stoechas* essential oil. Briefly, 1 mL of each test solution was applied to filter paper discs placed in Petri dishes (90 mm diameter) in such a way as to generate a homogenous film. The treated Petri dishes were first air-dried for a short time to let the solvent evaporate, then the larvae ($n = 10$) were released, and finally the Petri dishes were kept under observation in the laboratory for 24 h. The positive control group was treated with deltamethrin at a concentration of 2 L/mL, while the negative control group was treated with acetone. Three replicates of each test were performed.

$$\text{Percentage mortality} = \frac{\text{Number of dead larvae}}{\text{Number of larvae introduced}} \times 100$$

2.13. Pupicidal Bioassay against Musca domestica

In this bioassay, ten 2- to 3-day-old pupae were placed in a glass Petri dish. They then received a single application of 10 µL of each test solution [55]. Acetone was given to the negative control group, while the positive control group received deltamethrin at a concentration of 2 µL/mL. The Petri dishes were put in an incubator set to $28 \pm 2\ °C$ and 75–85% relative humidity. The treated pupae were monitored for six days to evaluate the emergence of adults. Three replicates of each test were performed. The adult emergence rate was evaluated following the method of Kumar et al. [56,57] and the percentage of inhibition rate (PIR) was calculated using the following equation:

$$PIR = \frac{\text{Number of newly emerged insect in control} - \text{Number of newly emerged insect in the treated}}{\text{Number of newly emerged insect in control}} \times 100$$

2.14. Statistics

For each treatment, three to five replicates were carried out and mean ± SE values were calculated. ANOVA was used to analyze larval mortality, followed by Duncan's multiple range test ($p < 0.05$). To determine the LC50 and LC90 values, as well as their 95% confidence limits, probit analyses were used [58]. SPSS for Windows (version 22.0) was used to conduct all statistical analyses.

3. Results

3.1. Chemical Composition of the Essential Oil

Hydrodistillation yielded a pale-yellow essential oil. The yield was 1.9% (v/w). Thirty-six compounds were identified, accounting for 95.56% of the EO's total volatile fraction. The GC–MS analyses of the *L. stoechas* essential oil showed that the predominant compound was camphor (58.38%), followed by fenchone (18.15%) and eucalyptol (6.93%). The fraction of oxygenated monoterpenes constituted almost 87% of the oil, while hydrocarbon monoterpenes and sesquiterpenes constituted about 4.8% and 3.4%, respectively (Table 1 and Figure 1).

Table 1. Chemical composition of the essential oil of *Lavandula stoechas*.

Peak No	RT (min)	R.I.$_{ex}$	R.I.$_{lt}$	COMPOUND	%	
1	8.33	930	939	α-Pinene	0.55	MH
2	9.14	945	954	Camphene	2.04	MH
3	9.36	960	960	Thuja-2,4(10)-diene	0.68	MH
4	10.38	987	990	β-Myrcene	0.22	MH
5	11.04	998	1130	2,6-Dimethyl-1,3, 5,7-octatetraene Unknown $^{\&}$	0.53	other
6	11.80	1021	1024	p-cymene	0.19	MH
7	11.99	1024	1029	Limonene	1.13	MH
8	12.09	1029	1031	Eucalyptol	6.93	OM
9	13.14	1057	1059	γ-Terpinene	Trace *	MH
11	13.68	1069	1072	cis-Linalool oxide (furanoid)	Trace *	OM
12	14.3	1083	1086	trans-Linalool oxide (furanoid)	Trace *	OM
13	14.69	1088	1086	Fenchone	18.15	OM
14	15.99	1122	1116 (endo) or 1121 (exo)	Fenchol **	Trace *	OM
15	16.29	1129	1126	α-Campholene aldehyde	Trace *	OM
16	17.40	1151	1146	Camphor	58.38	OM
17	18.46	1175	1169	Borneol	0.12	OM
18	18.69	1181	1177	Terpinen-4-ol	Trace *	OM
19	19.38	1195	1188	α-Terpineol	0.13	OM
21	19.53	1197	1195	Myrtenal	0.45	OM
22	20.10	1210	1205	Verbenone	0.33	OM
23		1214	1220	Fenchyl acetate **	Trace *	Mester
24	21.71	1248	1243	Carvone	0.17	OM
25	23.13	1279	1288	Bornyl acetate	1.40	Mester
26	24.81	1318	1326	Myrtenyl acetate	0.40	Mester
27	25.66	1338	1348	α-Cubebene	Trace *	HS
28	26.79	1363	1376	Isoledene?	0.23	HS
29	27.00	1368	1376	α-Copaene	0.10	HS
30	27.23	1373	1544	Isolongifolene-4,5, 9,10-dehydro?? Unknown $^{\&}$	0.37	HS
31	31.07	1465	1476	Cadina-1(6),4-diene, trans	0.21	HS
32	31.84	1484	1496	Viridiflorene (syn. Ledene)	0.93	HS
33	32.15	1490	1500	α-Muurolene	0.41	HS
34	32.35	1496	1517	α-dehydro Himachalene?	Trace *	HS
35	32.92	1509	1523	δ-Cadinene	0.74	HS
36	33.19	1516	1529	cis-Calamenene	0.21	HS
37	33.59	1527	1534	trans-Cadina-1,4-diene	0.24	HS
38	34.04	1538	1545	α-Calacorene	0.35	HS
39	36.99	1615	1623	α-Corocalene	0.10	HS
40	39.21	1673	1676	Cadalene	0.10	HS
				Traces *	0.36	
				Total (without traces = 0.36%)	95.79	

RT: Retention time. R.I.$_{ex}$: Experimental retention index calculated on Rt-5MS column. R.I.$_{lt}$: Retention index from the literature for relative columns. * Components representing less than 0.1%. ** Correct isomer not identified. Unknown $^{\&}$: main m/z is given (% relative intensity). Comp No 5: 91 (100), 119 (58), 77 (40), 134 (32). Comp No 30: 143 (100), 157 (80), 200 (55), 128 (40), 185 (34). MH: Monoterpene hydrocarbon. OM: Oxygenated monoterpene. HS: Hydrocarbon sesquiterpene.

3.2. Cytotoxicity Assessment

Cell viability was affected by the highest concentrations of lavender, whereby the concentrations of 0.25, 0.5, and 1.0% showed toxicity against 80% of HEK-293 (Figure 2).

Additionally, this compound was shown to be safe for normal cells up to a concentration of 0.0625%, at which point a strong cytotoxic effect appeared at around 0.125%. Lavender oil showed clear cytotoxic effects on the MCF-7 cell line, even at a low concentration of 0.0625%, causing cell death at a rate of 30%, while high concentrations caused significant cell death (Figure 3).

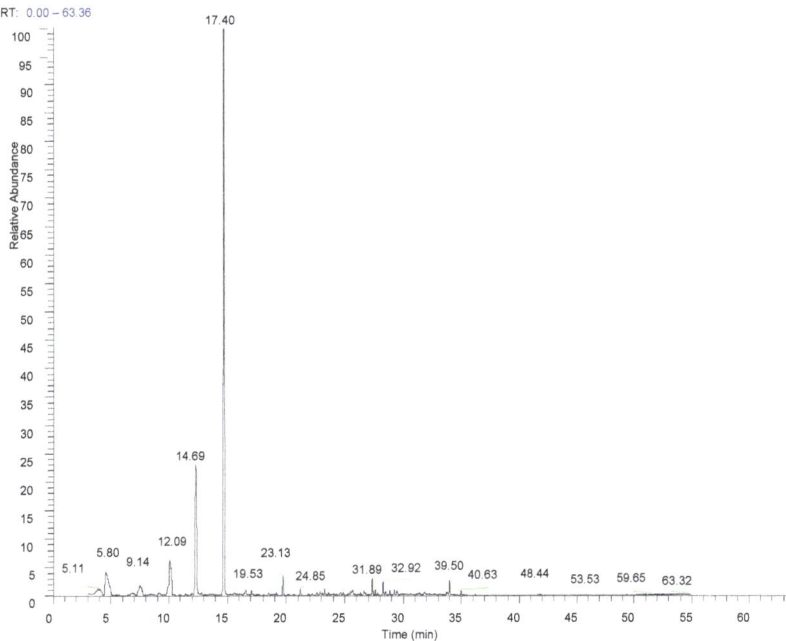

Figure 1. GC–MS chromatogram of *Lavandula stoechas* essential oil.

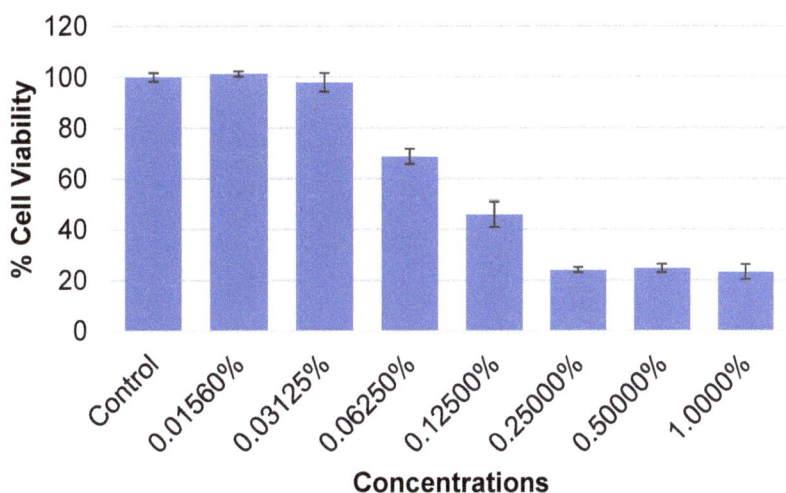

Figure 2. Cytotoxic activity of lavender oil against HEK-293 cells determined by MTT assay. Cells were exposed to different concentrations of the oil for 24 h. All values are presented as mean ± SD.

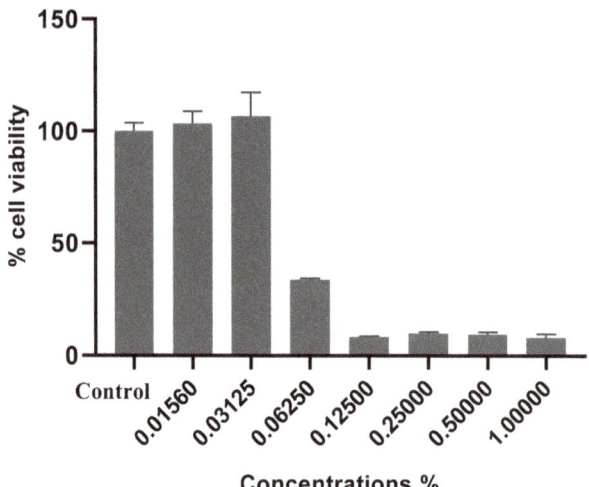

Figure 3. Cytotoxic activity of *L. steochas* EO against MCF-7 cells determined by MTT assay. Cells were exposed to different concentrations of EO for 24 h. All values are presented as mean ± SD.

3.3. In Vitro Scolicidal Activity

Table 2 displays the scolicidal effects of the *L. stoechas* essential oil at various concentrations and with different exposure times. At a concentration of 0.025%, the scolicidal effectivity values of *L. stoechas* oil were 33.66, 50.4, and 88.07% after 1, 3, and 5 min, respectively. The values at a concentration of 0.05% were 41.07, 72.47, and 98.17% after 1, 3, and 5 min respectively. After 3 and 5 min of exposure, an oil concentration of 0.1% caused 95.5% and 100% mortality, respectively (Figure 4). Overall, the scolicidal activity of the oil was clearly concentration- and time-dependent.

Table 2. Scolicidal effect of *Lavandula stoechas* essential oil on the viability of *E. granulosus* protoscolices.

Concentrations (%)	Mortality Rates after Exposure (%) (Mean ± SE)		
	1 min	3 min	5 min
0.025%	33.66 ± 1.52 [c]	50.4 ± 1.24 [c]	88.07 ± 1.33 [b]
0.05 %	41.07 ± 1.32 [b]	72.47 ± 1.76 [b]	98.17 ± 1.04 [a]
0.1 %	74.7 ± 1.99 [a]	95.5 ± 1.19 [a]	100.00 ± 0.00 [a]
Control	3.03 ± 0.29 [d]	4.1 ± 0.32 [d]	5.2 ± 0.23 [c]

Means within the same column followed by different superscripts are significantly different (Duncan's multiple range test: $p \leq 0.05$). LC = lethal concentration, CL = confidence limit, X^2 = chi square, df = degree of freedom.

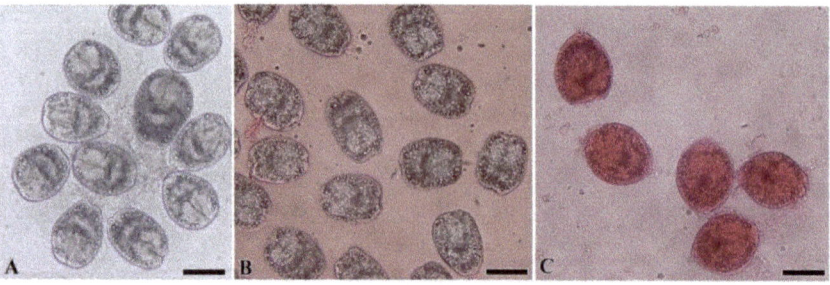

Figure 4. Live non-stained protoscolices (**A**), live protoscolices after staining with 0.1% eosin (**B**), dead protoscolices after treatment with *L. steochas* EO and staining with 0.1% eosin (**C**). Scale-bar = 100 mm.

3.4. Acaricidal Activity L. stoechas EO against Adult and Larvae of R. annulatus Ticks

L. stoechas EO showed significant adulticidal activity against *R. annulatus* ticks, especially at concentrations of 5 and 10%, showing tick mortality rates of 86.66 and 100%, respectively, and the LC_{50} was 2.34%. Moreover, the egg production index of the treated groups showed lower values compared to those of the control, i.e., the untreated ticks (Table 3).

Table 3. Adulticidal and lethal concentrations (LC_{50}, LC_{90}) of *Lavandula stoechas* against *R. annulatus* adult ticks.

Concentrations %	Mortality % M ± SE	Egg Production Index (EPI)	LC_{50} (95% CL)	LC_{90} (95% CL)	χ^2 (df = 3)	p
0.625	0.00 ± 0.00 [e]	35.83 ± 0.60 [b]				
1.25	16.66 ± 3.33 [d]	31.80 ± 1.80 [c]	2.34	5.00	3.861	0.277
2.5	56.67 ± 6.67 [c]	26.00 ± 0.58 [d]	(2.13–2.57)	(4.39–5.89)		
5	86.66 ± 3.33 [b]	24.33 ± 0.67 [d]				
10	100.00 ± 0.00 [a]	0.00 ± 0.00 [e]				
Deltamethrin 2 uL/mL	26.67 ± 3.33 [c]	23.33 ± 0.88 [d]	-	-	-	-
Ethyl alcohol 70%	0.00 ± 0.00 [e]	43.00 ± 1.00 [a]	-	-	-	-

Means within the same column followed by different superscripts are significantly different (Duncan's multiple range test: $p \leq 0.05$). LC = lethal concentration, CL = confidence limit, χ^2 = chi square, df = degree of freedom.

Regarding larval toxicity, *L. stoechas* oil achieved a larvicidal activity of 86.7% at the highest concentration (10%), with an LC_{50} of 9.11% (Table 4).

Table 4. Larvicidal activity and lethal concentrations (LC_{50}, LC_{90}) of *Lavandula stoechas*, against larvae of *R. annulatus*.

Concentrations %	Mortality % M ± SE	LC_{50} (95% CL)	LC_{90} (95% CL)	χ^2 (df = 3)	p
0.625	6.67 ± 1.67 [f]				
1.25	15.00 ± 2.89 [e]	3.82	15.53	5.43	0.143
2.5	35.00 ± 2.88 [c]	(3.32–4.45)	(11.94–22.07)		
5	51.66 ± 1.67 [b]				
10	86.67 ± 1.66 [a]				
Deltamethrin 2 uL/mL	22.66 ± 2.66 [d]	-	-	-	-
Ethyl alcohol 70%	5.33 ± 1.66 [f]	-	-	-	-

Means within the same column followed by different superscripts are significantly different (Duncan's multiple range test: $p \leq 0.05$). LC = lethal concentration, CL = confidence limit, χ^2 = chi square, df = degree of freedom.

Moreover, regarding the repellence activity of *L. stoechas*, we found a weak repellency in the first hour, even at the highest concentration of 10%, and no repellent effect was seen in the succeeding hours (S Table 1). There were no significant differences in results between the low concentrations and the control, treated with 70% ethyl alcohol.

3.5. Insecticidal Effect of L. stoechas against Larvae and Pupae of Musca domestica

The larval toxicity of *L. stoechas* oil, assessed via the residual film method, increased significantly with increasing concentrations. Within 24 h after application, the lavender-treated groups showed significant mortality at 5 and 10% concentrations, with rates of 93.33 and 100%, respectively, and an LC_{50} value of 1.79%. The treated larvae died within 24 h, with clear blackening of the cuticles. Moreover, in the deltamethrin-treated and negative control groups, no larvicidal activities were observed whatsoever (Table 5).

Regarding pupal toxicity, *L. stoechas* oil achieved a percentage inhibition rate (PIR), ranging from 13.33% to 100% at various concentrations after six days of application with an LC_{50} value of 1.51%. Moreover, a concentration of 10% *L. steochas* essential oil led to the complete inhibition of adult emergence, and the dead pupae displayed darker colors (Table 6).

Table 5. Larvicidal activity, and lethal concentrations (LC$_{50}$, LC$_{90}$) of *Lavandula stoechas*, against larvae of *Musca domestics*.

Concentrations %	Mortality % (Mean ± SE)	LC$_{50}$ (95% CL)	LC$_{90}$ (95% CL)	χ2 (df = 3)	p
0.625	6.67 ± 3.33 [d]				
1.25	30.00 ± 5.77 [c]	1.79	4.29	0.845	0.839
2.5	66.66 ± 3.33 [b]	(1.62–1.98)	(3.72–5.13)		
5	93.33 ± 3.33 [a]				
10	100.00 ± 0.00 [a]				
Deltamethrin 2 uL/mL	0.00 ± 0.00 [d]	-	-	-	-
Ethyl alcohol 70%	0.00 ± 0.00 [d]	-	-	-	-

Means within the same column followed by different superscripts are significantly different (Duncan's multiple range test: $p \leq 0.05$). LC = lethal concentration, CL = confidence limit, X^2 = chi square, df = degree of freedom.

Table 6. Percentage inhibition rate (PIR) of *Lavandula stoechas* against housefly pupae in contact toxicity assay.

Concentrations %	PIR % (Mean ± SE)	LC$_{50}$ (95% CL)	LC$_{90}$ (95% CL)	χ2 (df = 3)	p
0.625	13.33 ± 3.33 [d]				
1.25	36.66 ± 6.67 [c]	1.51	3.94	1.667	0.644
2.5	76.67 ± 3.33 [b]	(1.35–1.68)	(3.38–4.78)		
5	93.33 ± 6.66 [a]				
10	100.00 ± 0.00 [a]				
Deltamethrin 2 uL/mL	0.00 ± 0.00 [e]	-	-	-	-
Ethyl alcohol 70%	0.00 ± 0.00 [e]	-	-	-	-

Means within the same column followed by different superscripts are significantly different (Duncan's multiple range test: $p \leq 0.05$). LC = lethal concentration, CL = confidence limit, X^2 = chi square, df = degree of freedom.

4. Discussion

Synthetic chemicals have been widely used to control parasitic infections, but their indiscriminate and excessive usage has resulted in drug resistance, as well as detrimental effects to the environment and food supply [25,26,57]. Plant essential oils (and/or active components) can be used as natural alternatives or adjuncts to current therapies used against a variety of ectoparasites and endoparasites of medical/veterinary significance [52,58–61]. The use of essential oils as therapeutic agents is more affordable, effective, and safe [14,62]. As such, the present work was designed to investigate the in vitro acaricidal, insecticidal, and scolicidal activities of *L. steochas* essential oil against certain cell lines, as well as its safety.

Our GC–MS analyses revealed camphor and fenchone as the main components of *L. steochas* essential oil, accounting for 58.38% and 18.15%, respectively, followed by 6.93% eucalyptol and 2.04% camphene. These outcomes agree with those of several previous studies, which showed that the predominant components of *L. steochas* oil are camphor and fenchone [63–67]. The timing of plant collection, the duration of hydrodistillation, the selection of plant parts to be used, and environmental factors were all found to have a significant impact on the essential oil's yield and composition [68].

The in vitro cytotoxic activity of *L. steochas* EO against human embryonic kidney cells (HEK-293 cells) and human breast cancer cell line (MCF-7) MCF-7 cells was assessed in the present investigation. The results demonstrate that *L. steochas* oil was extremely cytotoxic to HEK-293 and MCF-7 cells, even at low concentrations. Our results corroborate those found in [69], who noted that *L. steochas* flowers were cytotoxic to Allium cepa root-tip meristem cells. According to Siddiqui et al. [70], the cytotoxic effect of *L. stoechas* EO, perhaps resulting from apoptosis, apparently induces deformations in the nuclei and cell membrane. The ethanolic fraction of *L. stoechas* has an anticancer effect, which may be attributed to the presence of phytosterols [70]. Furthermore, it strongly inhibits the growth of human gastric adenocarcinoma (AGS), melanoma MV3, and breast cancer MDA-MB-231 cells, with median inhibitory concentrations (IC50) of 0.035 ± 0.018, 0.06 ± 0.022, and 0.259 ± 0.089 µL/mL, respectively [71].

We assessed the in vitro scolicidal efficacy of *L. steochas* EO at a variety of concentrations and exposure times on the protoscoleces of hydatidosis. The findings show that all tested concentrations of *L. steochas* EO had significant scolicidal activities, with 98.17% and 100% mortality achieved at doses of 0.05% and 0.1%, respectively, at 5 min post-treatment. The results of the current investigation suggest that *L. steochas* is a potential natural source of novel protoscolicidal agents that could be used in hydatid cyst surgery.

The current study revealed that *L. steochas* EO induced 100% mortality in *R. annulatus* adult ticks at concentration of 10%, with an LC_{50} value of 2.34%. Regarding the larvicidal activity, 86.67% larval death was achieved at a concentration of 10% with an LC_{50} of 9.11%. Similarly, *L. stoechas* EO showed significant acaricidal activities against adults and larvae of the *Hyalomma suspense* tick [10]. Other species of lavender have acaricidal activities. *Lavandula angustifolia* is effective against *Rhipicephalus* (*Boophilus*) *annulatus* [11], and *Lavandula luisieri* has larvicidal effects against *Hyalomma lusitanicum* [12]. Additionally, Sertkaya et al. [13] found that *L. steochas* EO had an acaricidal effect against the red spider mite *Tetranychus cinnabarinus*. It was surprising, however, to see that *L. stoechas* essential oil has no repellent effect against larvae of *R. annulatus*. Other studies did identify a repellent effect of *L. stoechas* and other types of lavender. *Lavandula angustifolia* shows repellency against *Hyalomma marginatum* adults [72] and nymphs of *Ixodes ricinus* (L.) (Acari: Ixodidae) [73]. This discrepancy may be due to the tick species and tick stages used in our tests.

In terms of the insecticidal activity against *M. domestica*, our results indicate that *L. steochas* EO exhibits maximum efficacy (100%) against house fly larvae at a concentration of 10% with an LC_{50} of 1.79% at 24 h post-application. Additionally, a 10% concentration of *L. steochas* EO showed the highest level of toxicity against house fly pupae, and completely inhibited adult emergence (100% PIR). Essential oils from the genus *Lavandula* have also shown insecticidal efficacy against several insect species. The essential oil of *L. stoechas*, a member of this genus, showed significant toxic effects against the adults and/or larvae of *Anopheles labranchiae*, *Culex pipiens molestus*, and *Orgyia trigotephras* [33,67,74]. The essential oils of *L. dentate* and *L. angustifolia*, also belonging to the same genus, exhibited significant insecticidal actions against larvae of *M. domestica* and *Chrysoma albiceps* and have thus been suggested for use as a safe and effective natural means to control these dipterans [75–80]. Bosly [79] observed morphological abnormalities in *M. domestica* larvae after treatment with *Lavandula* spp. EOs and attributed these deformities to hormonal imbalances that interrupt insect metamorphosis. Meanwhile, Khater and Khater [81] found that the oil may limit larval motility and prevent the larvae from constricting during the pupal stage, thus contributing to the observed deformities. According to Conti et al. [31] and Sajfrtova et al. [82], the concentration of volatile compounds in the oil is directly related to the toxic effects found. It is difficult to compare the efficacy of essential oils across different studies because the methods used for oil extraction vary greatly, and this impacts the subsequent essential oils' efficacy [79].

Generally, the presence of volatile components in essential oils is largely responsible for the oils' acaricidal, insecticidal, and cytotoxic actions [80–82]. For instance, the main component of *L. stoechas* EO, camphor, has been proven to have insecticidal effects [83,84] and creates a fragrant vapor that repels mosquitoes [66,67]. Another key ingredient in *L. stoechas* EO, camphene, also has insecticidal activities resulting from its ability to repel insects, including flies and moths [63,64,83,84]. All of this can thus explain *L. stoechas*' effectiveness as an acaricidal, insecticidal, scolicidal, and anticancer agent.

In conclusion, *L. stoechas* EO has potential applicability as a potent protoscolicidal agent with high effectiveness at low doses and in shorter times. However, more research is required to fully assess the potential use of this oil in the prevention and treatment of cystic echinococcosis, including via in vivo assays and its main components. Additionally, this oil shows significant activity against *M. domestica* and *R. annulatus* ticks, which pose a substantial risk to public health; camphor is the first known substance to exhibit this activity. The mechanisms of action of these essential oils are poorly understood. One of the theories is that the monoterpenes operate on other sensitive locations, such as the neurological system, but further research is required to verify and expand on this.

Author Contributions: Conceptualization, A.-A.S.A.-B., S.M.A. and A.A.K.; methodology, A.A.K., A.S. and S.M.A.; software, D.D. and A.O.H.; validation, A.S., D.D. and A.A.K.; formal analysis, A.A.K. and A.S.; investigation, A.-A.S.A.-B., A.A.K. and S.M.A.; resources, S.A.-Q.; data curation, D.D. and A.S.; writing—original draft preparation, A.A.K. and S.M.A.; writing—review and editing, A.-A.S.A.-B., S.A.-Q. and S.M.A.; visualization, A.O.H. and S.A.-Q.; supervision, A.-A.S.A.-B. and S.M.A.; project administration, S.A.-Q.; funding acquisition, S.A.-Q. All authors have read and agreed to the published version of the manuscript.

Funding: This work was supported by the Researcher supporting Project [RSP-2021/3], King Saud University.

Institutional Review Board Statement: Not applicable.

Informed Consent Statement: Not applicable.

Data Availability Statement: All relevant data are within the paper and its supporting information.

Conflicts of Interest: The authors declare no conflict of interest.

References

1. Panigrahi, P.N.; Gupta, A.R.; Behera, S.K.; Panda, B.S.K.; Patra, R.C.; Mohanty, B.N.; Sahoo, G.R. Evaluation of gastrointestinal helminths in canine population of Bhubaneswar, Odisha, India: A public health appraisal. *Vet. World* **2014**, *7*, 295–298. [CrossRef]
2. Abdel-Baki, A.S.; Almalki, E.; Mansour, L.; Al-Quarishy, S. In vitro scolicidal effects of Salvadora persica root extract against protoscolices of Echinococcus granulosus. *Korean J. Parasitol.* **2016**, *54*, 61–66. [CrossRef] [PubMed]
3. Maurice, M.N.; Huseein, E.A.M.; Monib, M.E.M.M.; Alsharif, F.M.; Namazi, N.I.; Ahmad, A.A. Evaluation of the scolicidal activities of eugenol essential oil and its nanoemulsion against protoscoleces of hydatid cysts. *PLoS ONE* **2021**, *16*, e0259290. [CrossRef] [PubMed]
4. Rokni, M. Echinococcosis/hydatidosis in Iran. *Iran J. Parasitol.* **2009**, *4*, 1e16.
5. Norouzi, R.; Ataei, A.; Hejazy, M.; Noreddin, A.; El Zowalaty, M.E. Scolicidal Effects of Nanoparticles Against Hydatid Cyst Protoscolices in vitro. *Int. J. Nanomed.* **2020**, *15*, 1095–1100. [CrossRef] [PubMed]
6. Shakibaiea, M.; Khalaf, A.K.; Rashidipour, M.; Mahmoudvand, H. Effects of green synthesized zinc nanoparticles alone and along with albendazole against hydatid cyst protoscoleces. *Ann. Med. Surg.* **2022**, *78*, 10. [CrossRef]
7. Junghanss, T.; Da Silva, A.M.; Horton, J.; Chiodini, P.L.; Brunetti, E. Clinical management of cystic echinococcosis: State of the art, problems, and perspectives. *Am. J. Trop. Med. Hyg.* **2008**, *79*, 301e311. [CrossRef]
8. Al-Harbi, N.A.; Al Attar, N.M.; Hikal, D.M.; Mohamed, S.E.; Abdel Latef, A.A.H.; Ibrahim, A.A.; Abdein, M.A. Evaluation of Insecticidal Effects of Plants Essential Oils Extracted from Basil, Black Seeds and Lavender against Sitophilus oryzae. *Plants* **2021**, *10*, 829. [CrossRef]
9. Anthony, J.P.; Fyfe, L.; Smith, H. Plant active components–a resource for antiparasitic agents? *Trends Parasitol.* **2005**, *21*, 462–468. [CrossRef]
10. Brunetti, E.; Kern, P.; Vuitton, D.A. Writing Panel for the WHO-IW-GE. Expert consensus for the diagnosis and treatment of cystic and alveolar echinococcosis in humans. *Acta Trop.* **2010**, *114*, 1–16. [CrossRef]
11. Almalki, E.; Al-Shaebi, E.M.; Al-Quarishy, S.; El-Matbouli, M.; Abdel-Baki, A.S. In vitro effectiveness of *Curcuma longa* and *Zingiber officinale* extracts on *Echinococcus* protoscoleces. *Saudi J. Biol. Sci.* **2017**, *24*, 90–94. [CrossRef] [PubMed]
12. Moazeni, M.; Saharkhiz, M.J.; Hosseini, A.A. In vitro lethal effect of ajowan (*Trachyspermum ammi* L.) essential oil on hydatid cyst protoscoleces. *Vet. Parasitol.* **2012**, *187*, 203–208. [CrossRef] [PubMed]
13. Moazeni, M.; Nazer, A. In vitro effectiveness of garlic (*Allium sati-vum*) extract on scolices of hydatid cyst. *World J. Surg.* **2010**, *34*, 2677–2681. [CrossRef] [PubMed]
14. Abbas, A.; Abbas, R.Z.; Khan, J.A.; Iqbal, Z.; Bhatti, M.M.H.; Sindhu, Z.u.D.; Zia, M.A. Integrated strategies for the control and prevention of dengue vectors with particular reference to *Aedes aegypti*. *Pak. Vet. J.* **2014**, *34*, 1–10.
15. Yadav, P.K.; Rafiqi, S.M.; Panigrahi, P.N.; Kumar, D.; Kumar, R.; Kumar, S. Recent trends in control of ectoparasites: A review. *J. Entomol. Zool. Stud.* **2017**, *5*, 808–813.

16. Jongejan, F.; Uilenberg, G. The global importance of ticks. *Parasitology* **2004**, *129* (Suppl. 1), S3–S14. [CrossRef]
17. Chen, Z.; Liu, Q.; Liu, J.Q.; Xu, B.L.; Lv, S.; Xia, S.; Zhou, X.N. Tick-borne pathogens and associated co-infections in ticks collected from domestic animals in central China. *Parasit Vectors* **2014**, *7*, 1–8. [CrossRef]
18. Demessie, Y.; Derso, S. Tick borne hemoparasitic diseases of ruminants: A Review. *Adv. Biol. Res.* **2015**, *9*, 210–224.
19. Opara, M.N.; Santali, A.; Mohammed, B.R.; Jegede, O.C. Prevalence of haemoparasites of small ruminants in Lafia Nassarawa State: A Guinea Savannah Zone of Nigeria. *J. Vet. Adv.* **2016**, *6*, 1251–1257.
20. Kamaraj, C.; Rajakumar, G.; Rahuman, A.A.; Velayutham, K.; Bagavan, A.; Zahir, A.A.; Elango, G. Feeding deterrent activity of synthesized silver nanoparticles using *Manilkara zapota* leaf extract against the house fly, *Musca domestica* (Diptera: Muscidae). *Parasitol. Res.* **2012**, *111*, 2439–2448. [CrossRef]
21. Khan, H.A.; Akram, W.; Arshad, M.; Hafeez, F. Toxicity and resistance of field collected *Musca domestica* (Diptera: Muscidae) against insect growth regulator insecticides. *Parasitol. Res.* **2016**, *115*, 1385–1390. [CrossRef] [PubMed]
22. Morey, R.A.; Khandagle, A.J. Bioefficacy of essential oils of medicinal plants against housefly, *Musca domestica* L. *Parasitol. Res.* **2012**, *111*, 1799–1805. [CrossRef] [PubMed]
23. Sinthusiri, J.; Soonwera, M. Oviposition deterrent and ovicidal activities of seven herbal essential oils against female adults of housefly, *Musca domestica* L. *Parasitol. Res.* **2014**, *113*, 3015–3022. [CrossRef]
24. Foil, L.D.; Coleman, P.; Fragoso-Sanchez, H.E.; Garcia-Vazquez, Z.; Guerrero, F.D.; Jonsson, N.N.; Langstaff, I.G.; Li, A.Y.; Machila, N.; Miller, R.J.; et al. Factors that influence the prevalence of acaricide resistance and tick-borne diseases. *Vet. Parasitol.* **2004**, *125*, 163–181. [CrossRef] [PubMed]
25. El-Seedi, H.R.; Azeem, M.; Khalil, N.S.; Sakr, H.H.; Shaden, A.; Khalifa, M.; Awang, K.; Saeed, A.; Mohamed AFarag, M.A.; Al Ajmi, M.F.; et al. Essential oils of aromatic Egyptian plants repel nymphs of the tick *Ixodes ricinus* (Acari: Ixodidae). *Exp. Appl. Acarol.* **2017**, *73*, 139–157. [CrossRef]
26. Showler, A.T. Botanically based repellent and insecticidal effects against horn flies and stable flies (Diptera: Muscidae). *J. Integr. Pest. Manag.* **2017**, *8*, 1–11. [CrossRef]
27. Khan, M.N.; Sajid, M.S.; Rizwan, H.M.; Qudoos, A.; Abbas, R.Z.; Riaz, M.; Khan, M.K. Comparative efficacy of six anthelmintic treatments against natural infection of *fasciola* species in sheep. *Pak. Vet. J.* **2017**, *37*, 65–68.
28. Liaqat, I.; Pervaiz, Q.; Bukhsh, S.J.; Ahmed, S.I.; Jahan, N. (2016) Investigation of bactericidal effects of medicinal plant extracts on clinical isolates and monitoring their biofilm forming potential. *Pak. Vet. J.* **2016**, *36*, 159–164.
29. Chintalchere, J.M.; Dar, M.A.; Pandit, R.S. Biocontrol efficacy of bay essential oil against housefly, *Musca domestica* (Diptera: Muscidae). *J. Basic Appl. Zool.* **2020**, *81*, 1–12. [CrossRef]
30. El-Akhal, F.; El Ouali Lalami, A.; Ez Zoubi, Y.; Greche, H.; Guemmouh, R. Chemical composition and larvicidal activity of essential oil of *Origanum majorana* (Lamiaceae) cultivated in Morocco against *Culex pipiens* (Diptera:Culicidae). *Asian Pac. J. Trop. Biomed.* **2014**, *4*, 746–750. [CrossRef]
31. Conti, B.; Angelo, C.; Alessandra, B.; Francesca, G.; Luisa, P. Essential oil composition and larvicidal activity of six Mediterranean aromatic plants against the mosquito *Aedes albopictus* (Diptera: Culicidae). *Parasitol. Res.* **2010**, *107*, 1455–1461. [CrossRef] [PubMed]
32. Miller, A.G. The Genus *Lavandula* in Arabia and Tropical NE Africa. *Notes R. Bot. Gard. Edinb.* **1985**, *42*, 503–528.
33. Badreddine, B.S.; Olfa, E.; Samir, D.; Hnia, C.; Lahbib, B.J.M. Chemical composition of *Rosmarinus* and *Lavandula* essential oils and their insecticidal effects on *Orgyia trigotephras* (Lepidoptera, Lymantriidae). *Asian Pac. Trop. Med.* **2015**, *8*, 98–103. [CrossRef] [PubMed]
34. Gamez, M.J.; Jimenez, J.; Risco, S.; Zarzuelo, A. Hypoglycemic activity in various species of the genus *Lavandula*: *Lavandula stoechas* L. and *Lavandula multifida* L. *Pharmazie* **1987**, *10*, 706–707.
35. Cavanagh, H.M.A.; Wilkinson, J.M. Biological Activities of Lavender essential oil. *Phytother. Res.* **2002**, *16*, 301–308. [CrossRef]
36. Dadalioglu, I.; Evrendilek, G.A. Chemical compositions and antibacterial effects of essential oils of turkish oregano (*Origanum minutiflorum*), bay laurel (*Laurus nobilis*), spanish lavender (*Lavandula stoechas*), and fennel (*Foeniculum vulgare*) on common foodborne pathogens. *J. Agri. Food Chem.* **2004**, *52*, 8255–8260. [CrossRef]
37. Benabdelkader, T.; Zitouni, A.; Guitton, Y.; Jullien, F.; Maitre, D.; Casabianca, H.; Legendre, L.; Kameli, A. Essential oils from wild populations of Algerian *Lavandula stoechas* L.: Composition, chemical variability, and in vitro biological properties. *Chem. Biodivers.* **2011**, *8*, 937–953. [CrossRef]
38. Angioni, A.; Barra, A.; Coroneo, V.; Desi, S.; Cabras, P. Chemical composition, seasonal variability and antifungal activity of *Lavandula stoechas* L. ssp. Stoechas essential oils from stem/leaves and flowers. *J. Agri. Food. Chem.* **2006**, *54*, 4364–4370.
39. Messaoud, C.; Chongrani, H.; Boussaid, M. Chemical composition and antioxidant activities of essential oils and methanol extracts of three wild *Lavandula*, L. species. *Nat. Prod. Res.* **2012**, *26*, 1976–1984. [CrossRef]
40. Ezzoubi, Y.; Bousta, D.; Lachkar, M.; Farah, A. Antioxidant and anti-inflammatory properties of ethanolic extract of *Lavandula stoechas* L. from Taounate region in Morocco. *Int. J. Phytopharm.* **2014**, *5*, 21–26.
41. Sokmen, A.; Abdel-Baki, A.A.S.; Al-Malki, E.S.; Al-Quraishy, S.; Abdel-Haleem, H.M. Constituents of essential oil of *Origanum minutiflorum* and its in vitro antioxidant, scolicidal and anticancer activities. *J. King Saud Univ. Sci.* **2020**, *32*, 2377–2382. [CrossRef]
42. Adams, R.P. *Identification of Essential Oil Components by Gas Chromatography/Mass Spectroscopy*, 4th ed.; Allured Publishing Corporation: Carol Stream, IL, USA, 2007.

43. Siddiqui, M.A.; Singh, G.; Kashyap, M.P.; Khanna, V.K.; Yadav, S.; Chandra, D.; Pant, A.B. Influence of cytotoxic doses of 4-hydroxynonenal on selected neurotransmitter receptors in PC-12 cells. *Toxicol. Vitr.* **2008**, *22*, 1681–1688. [CrossRef]
44. Haghani, A.; Roozitalab, A.; Safi, S.N. Low scolicidal effect of *Ocimum bacilicum* and *Allium cepa* on protoccoleces of hydatid cyst: An in vitro study. *Comp. Clin. Pathol.* **2014**, *23*, 847–853. [CrossRef]
45. Estrada-Peña, A.; Quı́ez, J.; Sánchez Acedo, C. Species composition, distribution, and ecological preferences of the ticks of grazing sheep in north-central Spain. *Med. Vet. Entomol.* **2004**, *18*, 123–133. [CrossRef] [PubMed]
46. Drummond, R.; Ernst, S.; Trevino, J.; Gladney, W.; Graham, O. (1973) *Boophilus annulatus* and *B. microplus*: Laboratory tests of insecticides. *J. Econ. Entomol.* **1973**, *66*, 130–133. [CrossRef] [PubMed]
47. FAO. *Resistance Management and Integrated Parasite Control in Ruminants Guidelines, Module Ticks: Acaricide Resistance: Diagnosis, Management and Prevention*; Food and Agriculture Organization, Animal Production and Health Division: Rome, Italy, 2004.
48. Klafke, G.M.; Thomas, D.B.; Miller, R.J.; Pérez de León, A.A. Efficacy of a water-based botanical acaricide formulation applied in portable spray box against the southern cattle tick, *Rhipicephalus (Boophilus) microplus* (Acari: Ixodidae), infesting cattle. *Ticks Tick Borne Dis.* **2021**, *12*, 101721. [CrossRef] [PubMed]
49. Matos, R.S.; Daemon, E.; de Oliveira Monteiro, C.M.; Sampieri, B.R.; Marchesini, P.B.C.; Delmonte, C. (2019) Thymol action on cells and tissues of the synganglia and salivary glands of *Rhipicephalus sanguineus* sensu lato females (Acari: Ixodidae). *Ticks Tick Borne Dis.* **2019**, *10*, 314–320. [CrossRef] [PubMed]
50. Wanzala, W.; Sika, N.; Gule, S.; Hassanali, A. Attractive and repellent host odours guide ticks to their respective feeding sites. *Chemoecology* **2004**, *14*, 229–232. [CrossRef]
51. Jesikha, M. Control of *Musca domestica* using wastes from *Citrus sinensis* peel and *Mangifera indica* seed. *Biol. Environ. Sci.* **2014**, *1*, 17–26.
52. Abdel-Baki, A.S.; Aboelhadid, S.M.; Sokmen, A.; Al-Quraishy, S.; Hassan, A.O.; Kamel, A.A. (2021) Larvicidal and pupicidal activities of *Foeniculum vulgare* essential oil, trans-anethole and fenchone against house fly *Musca domestica* and their inhibitory effect on acetylcholinestrase. *Entomol. Res.* **2021**, *51*, 568–577. [CrossRef]
53. Busvine, J.R. *A Critical Review of the Techniques for Testing Insecticides*; CABI: London, UK, 1971; p. 345.
54. Palacios, S.M.; Bertoni, A.; Rossi, Y.; Santander, R.; Urzua, A. Efficacy of essential oils from edible plants as insecticides against the house fly, *Musca domestica* L. *Molecules* **2009**, *14*, 1938–1947. [CrossRef] [PubMed]
55. Sinthusiri, J.; Soonwera, M. Effect of Herbal Essential Oils against Larvae, Pupae and Adults of House Fly (Musca domestica L.: Diptera). In Proceedings of the 16th Asian Agricultural Symposium and 1st International Symposium on Agricultural Technology, Bangkok, Thailand, 25–27 August 2010; pp. 639–642.
56. Kumar, P.; Mishra, S.; Malik, A.; Satya, S. Repellent, larvicidal and pupicidal properties of essential oils and their formulations against the housefly, *Musca domestica*. *Med. Vet. Entomol.* **2011**, *25*, 302–310. [CrossRef] [PubMed]
57. Finney, D.J. *Probit Analysis: A Statistical Treatment of the Sigmoid Response Curve*, 2nd ed.; Cambridge University Press: Cambridge, UK, 1952.
58. Ellse, L.; Wall, R. The use of essential oils in veterinary ectoparasite control: A review. *Med. Vet. Entomol.* **2014**, *28*, 233–243. [CrossRef] [PubMed]
59. Aslam, A.; Shahzad, M.I.; Parveen, S.; Ashraf, H.; Naz, N.; Zehra, S.S.; Kamran, Z.; Qayyum, A.; Mukhtar, M. Evaluation of antiviral potential of different Cholistani plants against infectious bursal disease and infectious bronchitis virus. *Pak. Vet. J.* **2016**, *36*, 302–306.
60. Radsetoulalova, I.; Hubert, J.; Lichovnikova, M. Acaricidal activity of plant essential oils against poultry red mite (*Dermanyssus gallinae*). *MendelNet* **2017**, *24*, 260–265.
61. Aboelhadid, S.M.; Arafa, W.M.; Abdel-Baki, A.A.S.; Sokmen, A.; Al-Quraishy, S.; Hassan, A.O.; Kamel, A.A. Acaricidal activity of *Foeniculum vulgare* against *Rhipicephalus annulatus* is mainly dependent on its constituent from trans-anethone. *PLoS ONE* **2021**, *16*, e0260172. [CrossRef]
62. Esmaily, M.; Bandani, A.; Zibaee, I.; Sharijian, J.; Zare, S. Sublethal effects of *Artemisia annua* L and *Rosmatinus officinalis* L, essential oils on life table parameters of *Tetranychus urticae* (Acari: Tetranychidae). *Persian J. Acarol.* **2017**, *6*, 39–52.
63. Arabi, F.; Moharramipour, S.; Sefidkon, F. Chemical composition and insecticidal activity of essential oil from *Perovskia abrotanoides* (Lamiaceae) against *Sitophilus oryzae* (Coleoptera: Curculionidae) and *Tribolium castaneum* (Coleoptera: Tenebrionidae). *Int. J. Trop. Insect Sci.* **2008**, *28*, 144–150. [CrossRef]
64. Yeh, R.Y.; Shiu, Y.L.; Shei, S.C.; Cheng, S.C.; Huang, S.Y.; Lin, J.C.; Liu, C.H. Evaluation of the antibacterial activity of leaf and twig extracts of stout camphor tree, *Cinnamomum kanehirae*, and the effects on immunity and disease resistance of white shrimp, *Litopenaeus vannamei*. *Fish Shellfish. Immunol.* **2009**, *27*, 26–32. [CrossRef]
65. Govindarajan, M. Larvicidal and repellent properties of some essential oils against *Culex tritaeniorhynchus* Giles and *Anopheles subpictus* Grassi (Diptera: Culicidae). *Asian Pac. J. Trop. Med.* **2011**, *4*, 106–111. [CrossRef]
66. Slimane, B.B.; Ezzine, O.; Dhahri, S.; Ben Jamaa, M.L. Essential oils from two Eucalyptus from Tunisia and their insecticidal action on *Orgyia trigotephras* (Lepidotera, Lymantriidae). *Biol. Res.* **2014**, *47*, 29. [CrossRef] [PubMed]
67. El Ouali Lalami, A.; El-Akhal, F.; El Amri, N.; Maniar, S.; Faraj, C. State resistance of the mosquito *Culex pipiens* towards temephos central Morocco. *Bull. Soc. Pathol. Exot.* **2014**, *107*, 194–198. [CrossRef] [PubMed]
68. Toker, R.; Gölükcü; M; Tokgöz, H. Effects of distillation times on essential oil compositions of *Origanum minutiflorum* O. Schwarz Et. and P.H Davis. *J. Essent. Oil Res.* **2017**, *29*, 30–335. [CrossRef]

69. Çelik, T.; Aslantürk, Ö. Cytotoxic and genotoxic effects of *Lavandula stoechas* aqueous extracts. *Biologia* **2007**, *62*, 292–296. [CrossRef]
70. Siddiqui, M.; Siddiqui, H.H.; Mishra, A.; Usmani, A. Evaluation of Cytotoxic Activity of *Lavandula stoechas* Aerial Parts Fractions against HepG2 Cell Lines. *Curr. Bioact. Compd.* **2020**, *16*, 1281–1289. [CrossRef]
71. Boukhatem, M.N.; Sudha, T.; Darwish, N.H.E.; Chader, H.; Belkadi, A.; Rajabi, M.; Houche, A.; Benkebailli, F.; Oudjida, F.; Mousa, S.A. A New Eucalyptol-Rich Lavender (*Lavandula stoechas* L.) Essential Oil: Emerging Potential for Therapy against Inflammation and Cancer. *Molecules* **2020**, *25*, 3671. [CrossRef]
72. Djebir, S.; Ksouri, S.; Trigui, M.; Tounsi, S.; Boumaaza, A.; Hadef, Y.; Benakhla, A. Chemical Composition and Acaricidal Activity of the Essential Oils of Some Plant Species of Lamiaceae and Myrtaceae against the Vector of Tropical Bovine Theileriosis: *Hyalomma scupense* (syn. Hyalomma detritum). *BioMed Res. Int.* **2019**, *2019*, 7805467. [CrossRef]
73. Pirali-Kheirabadi, K.; Teixeira da Silva, J.A. *Lavandula angustifolia* essential oil as a novel and promising natural candidate for tick (*Rhipicephalus* (*Boophilus*) *annulatus*) control. *Exp. Parasitol.* **2010**, *126*, 184–186. [CrossRef]
74. Julio, L.F.; Díaz, C.E.; Aissani, N.; Valcarcel, F.; Burillo, J.; Olmeda, S.; González-Coloma, A. Ixodicidal compounds from pre-domesticated ndula luisieri. *Ind. Crops Prod.* **2017**, *110*, 83–87. [CrossRef]
75. Sertkaya, E.; Kaya, K.; Soylu, S. Acaricidal activities of the essential oils from several medicinal plants against the carmine spider mite (*Tetranychus cinnabarinus* Boisd.) (Acarina: Tetranychidae). *Ind. Crops Prod.* **2010**, *31*, 107–112. [CrossRef]
76. Mkolo, M.; Magano, S. Repellent effects of the essential oil of *Lavendula angustifolia* against adults of *Hyalomma marginatum rufipes*. *J. South Afr. Vet. Assoc.* **2007**, *78*, 149–152. [CrossRef]
77. Jaenson, T.G.T.; Garboui, S.; Palsson, K. Repellency of oils of lemon eucalyptus, geranium, and lavender and the mosquito repellent MyggA natural to *Ixodes ricinus* (Acari: Ixodidae) in the laboratory and field. *J. Med. Entomol.* **2006**, *43*, 731–736. [CrossRef] [PubMed]
78. Traboulsi, A.F.; Taoubi, K.; El-Haj, S.; Bessiere, J.M.; Rammal, S. Insecticidal properties of essential plant oils against the mosquito *Culex pipiens* molestus (Diptera:Culicidae). *Pest. Manag. Sci.* **2002**, *58*, 491–495. [CrossRef] [PubMed]
79. Bosly, B. Evaluation of insecticidal activities of *Mentha piperita* and *Lavandula angustifolia* essential oils against house fly, *Musca domestica* L. (Diptera: Muscidae). *J. Entomol. Nematol.* **2013**, *5*, 50–54. [CrossRef]
80. Cossetin, L.F.; Santi, E.M.T.; Cossetin, J.F.; Dillmann, J.B.; Baldissera, M.D.; Garlet, Q.I.; de Souza, T.P.; Loebens, L.; Heinzmann, B.M.; Machado, M.M.; et al. In vitro Safety and Efficacy of Lavender Essential Oil (Lamiales: Lamiaceae) as an Insecticide Against Houseflies (Diptera: Muscidae) and Blowflies (Diptera: Calliphoridae). *J. Econ. Entomol.* **2018**, *111*, 1974–1982. [CrossRef]
81. Khater, H.F.; Khater, D.F. The insecticidal activity of four medicinal plants against the blowfly *Lucilia sericata* (Diptera: Calliphoridae). *Int. J. Dermatol.* **2009**, *48*, 492–497. [CrossRef]
82. Sajfrtova, M.; Sovova, H.; Karban, J.; Rochova, K.; Pavela, R.; Barnet, M. Effect of separation method on chemical composition and insecticidal activity of Lamiaceae isolates. *Ind. Crops Prod.* **2013**, *47*, 69–77. [CrossRef]
83. Selles, S.M.A.; Kouidri, M.; González, M.G.; González, J.; Sánchez, M.; González-Coloma, A.; Sanchis, J.; Elhachimi, L.; Olmeda, A.S.; Tercero, J.M.; et al. Acaricidal and Repellent Effects of Essential Oils against Ticks: A. Review. *Pathogens* **2021**, *10*, 1379. [CrossRef]
84. George, D.R.; Sparagano, O.A.E.; Port, G.; Okello, E.; Shiel, R.S.; Guy, J.H. Repellence of plant essential oils to *Dermanyssus gallinae* and toxicity to the non-target invertebrate Tenebrio molitor. *Vet. Parasitol.* **2009**, *162*, 129–134. [CrossRef]

Disclaimer/Publisher's Note: The statements, opinions and data contained in all publications are solely those of the individual author(s) and contributor(s) and not of MDPI and/or the editor(s). MDPI and/or the editor(s) disclaim responsibility for any injury to people or property resulting from any ideas, methods, instructions or products referred to in the content.

Article

Greener Stability-Indicating HPLC Approach for the Determination of Curcumin in In-House Developed Nanoemulsion and *Curcuma longa* L. Extract

Nazrul Haq [1], Faiyaz Shakeel [1], Mohammed M. Ghoneim [2], Syed Mohammed Basheeruddin Asdaq [2], Prawez Alam [3], Saleh A. Alanazi [4,5] and Sultan Alshehri [6,*]

[1] Department of Pharmaceutics, College of Pharmacy, King Saud University, Riyadh 11451, Saudi Arabia
[2] Department of Pharmacy Practice, College of Pharmacy, AlMaarefa University, Ad Diriyah 13713, Saudi Arabia
[3] Department of Pharmacognosy, College of Pharmacy, Prince Sattam Bin Abdulaziz University, Al-Kharj 11942, Saudi Arabia
[4] King Abdullah International Medical Research Center, College of Pharmacy, King Saud Bin Abdulaziz University for Health Sciences, Riyadh 11481, Saudi Arabia
[5] Pharmaceutical Care Department, King Abdulaziz Medical City, Ministry of National Guard Health Affairs, Riyadh 11426, Saudi Arabia
[6] Department of Pharmaceutical Sciences, College of Pharmacy, AlMaarefa University, Ad Diriyah 13713, Saudi Arabia
* Correspondence: sshehri.c@mcst.edu.sa

Abstract: Despite the fact that several analytical methodologies have been reported for the determination of curcumin (CCM) in a wide range of sample matrices, the greener liquid chromatographic approaches to determine CCM are scarce in the literature. Therefore, this research is designed to develop and validate a greener stability-indicating "high-performance liquid chromatography (HPLC)" methodology to determine CCM in an in-house developed nanoemulsion, *Curcuma longa* L. extract, and commercial tablets. CCM was measured on a Nucleodur (150 mm × 4.6 mm) RP C_{18} column with 5 μm-sized particles. Ethanol and ethyl acetate (83:17 *v/v*) made up the greener eluent system, which was pumped at a flow speed of 1.0 mL/min. At a wavelength of 425 nm, CCM was detected. The greener HPLC methodology was linear in the 1–100 μg/mL range, with a determination coefficient of 0.9983. The greener HPLC methodology for CCM estimation was also rapid (R_t = 3.57 min), accurate (%recoveries = 98.90–101.85), precise (%CV = 0.90–1.11), and sensitive (LOD = 0.39 μg/mL and LOQ = 1.17 μg/mL). The AGREE approach predicted the AGREE score of 0.81 for the established HPLC technique, indicating an outstanding greenness profile. The utility of the greener HPLC methodology was demonstrated by determining CCM in the in-house developed nanoemulsion, *Curcuma longa* extract, and commercial tablets. The % amount of CCM in the in-house developed nanoemulsion, *Curcuma longa* extract, and commercial tablets was found to be 101.24%, 81.15%, and 78.41%, respectively. The greener HPLC methodology was able to detect its degradation product under various stress conditions, suggesting its stability-indication characteristics. These results suggested that CCM in developed nanoemulsion, plant extract samples, and commercial tablets may be routinely determined using the greener HPLC methodology.

Keywords: AGREE; *Curcuma longa*; curcumin; nanoemulsion; greener HPLC; validation

1. Introduction

Turmeric is a spice, which is obtained from *Curcuma longa* L [1]. It is a rich source of phenolic compounds, namely curcuminoids [1,2]. Three main curcuminoids have been reported in *C. longa* extract: curcumin (CCM) (Figure 1), demethoxycurcumin (DCCM), and bisdemethoxycurcumin (BDCCM) [1,3]. Marketed CCM contains CCM, DCCM, and BDCCM, but its main constituent is CCM [4]. Various therapeutic activities have been reported for *C. longa* extract, which are mainly due to the presence of curcuminoids [1,4].

Citation: Haq, N.; Shakeel, F.; Ghoneim, M.M.; Asdaq, S.M.B.; Alam, P.; Alanazi, S.A.; Alshehri, S. Greener Stability-Indicating HPLC Approach for the Determination of Curcumin in In-House Developed Nanoemulsion and *Curcuma longa* L. Extract. *Separations* **2023**, *10*, 98. https://doi.org/10.3390/separations10020098

Academic Editor: Josef Cvačka

Received: 11 December 2022
Revised: 24 January 2023
Accepted: 28 January 2023
Published: 1 February 2023

Copyright: © 2023 by the authors. Licensee MDPI, Basel, Switzerland. This article is an open access article distributed under the terms and conditions of the Creative Commons Attribution (CC BY) license (https://creativecommons.org/licenses/by/4.0/).

CCM is a yellow-colored pigment, which is used for the coloring of various food products [4]. In the literature, the variety of therapeutic activities of CCM are reported, including antioxidant [1,2], anti-inflammatory [5], antimicrobial [6], antibacterial [7], antiviral [8], antiparasitic [9], antimutagenic [10], and antiproliferative activities [11–13] etc. As a consequence, the quality control and standardization of CCM in its pharmaceutical products, food products, and herbal extracts are significant due to its variety of therapeutic activities.

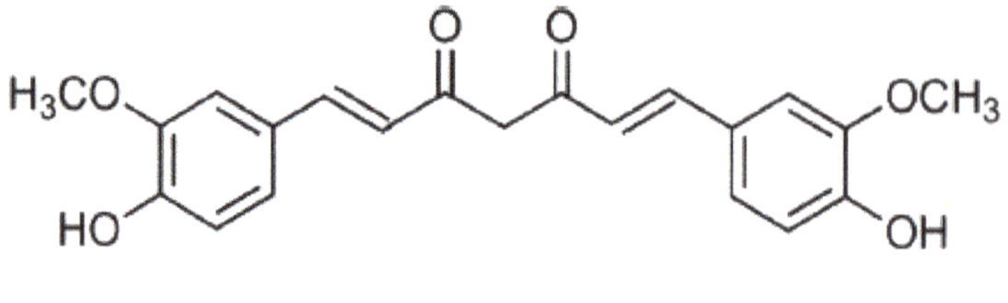

Figure 1. Molecular structure of curcumin (CCM).

Different analytical techniques have been utilized for the qualitative and quantitative detection of CCM in herbal extracts, commercial pharmaceutical formulations, commercial food products, and biological materials. For the determination of CCM in food products, spices, and herbal extracts, various ultra-violet (UV) spectrometry methods have been reported [14–16]. A spectrofluorometric assay has also been reported to determine CCM in a nanoliposomal formulation and mice plasma samples [17]. Fluorescence detection of CCM has also been carried out in the literature [18,19]. Reported spectrometry and fluorescence methods of CCM measurement were less accurate and sensitive than the current method [14–16,18,19]. The variety of high-performance liquid chromatographic methods (HPLC) have also been used to determine CCM in food products, pharmaceutical products, and herbal extracts [4,20–27]. However, most of the reported HPLC methods were environmentally toxic and less sensitive than the current HPLC method [4,20,25–27]. CCM in food products and herbal extracts has also been identified using liquid chromatography tandem mass-spectrometry (LC-MS) and ultra-performance liquid chromatography tandem mass-spectrometry (UPLC-MS) approaches [28–31]. Various high-performance thin-layer chromatography (HPTLC) approaches are also reported to determine CCM in herbal extracts and CCM dosage forms [32–37]. Some LC-MS and UPLC-MS approaches have also been used to determine CCM in the plasma samples of rat, equine, and human [38–41]. Carbon nanotube-based composites have also been used to determine CCM in food products [42,43]. Reported LC-MS, UPLC-MS, and HPTLC approaches were also found to be more environmentally toxic than the current HPLC method [28–43]. Some voltammetry approaches have also been reported to determine CCM in natural food supplements and food spices [44–46]. Some electrochemical approaches were also used to determine CCM in its pure form and human blood serum sample [47,48]. Some other techniques, such as the Fourier transform near infrared spectroscopy approach, nanosensor, and solvatochromic approach have also been reported to determine CCM [49–51].

The detailed literature survey revealed the wide range of analytical techniques for the determination of CCM in distinct sample matrices. However, the greener/sustainable HPLC approaches of CCM detection are scarce in the literature. Furthermore, no greenness index of any reported HPLC method of CCM analysis has been reported. Several qualitative and quantitative approaches are reported to evaluate the greener characteristics of analytical procedures [52–56]. However, the "analytical GREEnness (AGREE)" metric methodology exclusively considers all twelve green analytical chemistry (GAC) principles for the determination of the greenness profile [54]. As a result, the "AGREE metric methodology" was used for the determination of the greener profile of the present HPLC assay of CCM analysis [54]. Based on all these assumptions, the objective of this research was to design and validate a simple, rapid, and greener HPLC methodology for the determination of CCM in an in-house developed CCM nanoemulsion, *C. longa* extract, and commercial tablets. The greener HPLC methodology for determining CCM was validated, following the "International Council for Harmonization (ICH)-Q2-R1" procedures [57].

2. Materials and Methods

2.1. Materials

The working standard of CCM was obtained from "E-Merck (Darmstadt, Germany)". HPLC-grade solvents, such as methanol, ethyl acetate, acetone, and ethanol were provided by "Sigma Aldrich (St. Louis, MO, USA)". High-pure water was collected from "Milli-Q® water purifier (Millipore, Lyon, France)". All other chemicals and reagents used were of analytical grades. The fresh rhizomes of *Curcuma longa* were purchased from the hypermarket in Riyadh, Saudi Arabia. The commercial tablets (having 1000 mg of standardized *C. longa* extract) were obtained from a pharmacy shop in Riyadh, Saudi Arabia. The nanoemulsion formulation of CCM was developed in the laboratory by the aqueous phase titration method using clove oil (oil phase), Tween-20 (surfactant), Transcutol-HP (cosurfactant), and purified water (aqueous phase).

2.2. Instrumentation and Analytical Conditions

Waters HPLC system, composed of a 1515 isocratic pump, a 717 automatic sampler, quad LC-10A VP pumps, a programmable UV-visible variable wavelength detector, a column oven, an SCL 10AVP system controller, and an inline vacuum degasser, was used to measure CCM at a temperature of 25 ± 1 °C. The Millennium program (Version 32, Waters, Milford, MA, USA) was used to process and evaluate the data. CCM was determined using a Nucleodur (150 mm × 4.6 mm) RP C_{18} column with 5 µm-sized particles. The mixture of ethanol and ethyl acetate (83:17% v/v) was used as the greener eluent system. The eluent system flowed with a flow speed of 1.0 mL/min. At a wavelength of 425 nm, CCM was detected. The samples (20 µL) were injected into the system using a waters autosampler.

2.3. CCM Calibration Curve

In triplicates (n = 3), the appropriate amount of CCM was dispensed in the eco-friendly eluent system to produce a stock solution with a 200 µg/mL concentration. To obtain the serial dilutions in the required range (1–100 µg/mL) in triplicates (n = 3), the requisite aliquots from the stock solution of CCM (200 µg/mL) were diluted with the greener eluent system. Using the greener HPLC methodology, the chromatographic response for each concentration of CCM was identified. To produce the CCM calibration curve, CCM concentrations were plotted against the measured chromatographic response in triplicates (n =3).

2.4. Sample Preparation for the Determination of CCM in Curcuma Longa Extract

Approximately 10.0 g of fresh rhizomes of *C. longa* were taken and powdered finely. The fine powder was soaked in 100 mL of ethanol-water mixture (50:50 v/v) and ultrasonicated for about 60 min. The temperature was kept constant at 25 °C. The supernatant was recovered and filtered using nylon filter paper. The solvents were evaporated using a rotary vacuum evaporator at 40 °C. The obtained sample was used to determine CCM in *C. longa* extract using the greener HPLC approach.

2.5. Analytical Method Development

As the eluent systems, different combinations of greener solvents were examined to develop a trustworthy stability-indicating greener HPLC assay for the detection of CCM in an in-house developed nanoemulsion, *C. longa* extract, and commercial tablets. The greener solvent compositions of methanol-water, ethanol-water, acetone-water, methanol-ethanol, ethyl acetate-methanol, ethyl acetate-ethanol, ethyl acetate-acetone, acetone-ethanol, and acetone-methanol were among the numerous greener solvents that were examined. Various aspects were taken into account when determining the best greener eluent system, including the affordability, greenness/toxicity profile, the assay's sensitivity, the analysis duration, the chromatographic parameters, and the solvents' compatibilities with one another. As a consequence, various greener solvent compositions were examined as the eluent system in combined forms. Finally, the most trustworthy eluent system for future investigation was discovered to be a 83:17 volume-to-volume blend of ethanol and ethyl acetate.

2.6. Validation Parameters

Following ICH-Q2-R1 procedures, the greener HPLC technique for the measurement of CCM was verified for various parameters [57]. By drawing the linearity graphs, the linearity of the greener HPLC methodology was examined in the 1–100 µg/mL range. CCM solutions that had just been prepared were added to the HPLC system in triplicates ($n = 3$), and the chromatographic response was recorded. The CCM calibration curve was derived by plotting the CCM concentration vs. chromatographic response.

The system appropriateness parameters for the greener HPLC technology were derived using a number of chromatographic characteristics, including resolution (Rs), selectivity factor (α), tailing factor (As), capacity factor (k), and theoretical plates number (N) [58,59].

The intra-day and inter-day accuracy of the greener HPLC technology was assessed using a standard addition/spiking method in terms of % recovery [57]. To investigate intra-day accuracy at three different quality control (QC) levels, the pre-analyzed QC level of CCM (10 µg/mL) was spiked with an extra 50, 100, 150% of CCM solution to obtain low QC (LQC = 15 µg/mL), middle QC (MQC = 20 µg/mL), and high QC (HQC = 25 µg/mL). The obtained LQC, MQC, and HQC of CCM were analyzed on the same day in three replicates ($n = 3$) to measure intra-day accuracy. On three distinct days, three replicates ($n = 3$) of CCM's LQC, MQC, and HQC levels were obtained by the spiking method and used to examine inter-day accuracy. The percentage recovery, percentage coefficient of variance (%CV), and standard error were computed at each QC level.

The greener HPLC methodology's precision was examined using intra-day and inter-day variations. On the same day, three replicates ($n = 3$) of the LQC, MQC, and HQC levels of CCM were used to assess the intra-day precision. At the LQC, MQC, and HQC of the CCM on three distinct days, inter-day precision was assessed in three replicates ($n = 3$).

To examine the influence of intentional chromatographic alterations on CCM measurement, the robustness of the greener HPLC methodology was evaluated. The CCM MQC (20 µg/mL) was chosen for the robustness assessment. By altering the greener eluent's composition, flow speed, and detecting wavelength, robustness was assessed. The initial ethanol: ethyl acetate (83:17 *v/v*) eluent system was altered to ethanol: ethyl acetate (85:15 *v/v*) and ethanol: ethyl acetate (81:19 *v/v*) for the robustness assessment, and the differences in chromatographic response were recorded. For the purpose of evaluating robustness, the original flow speed of 1 mL/min was changed to flow rates of 1.15 mL/min and 0.85 mL/min, and the changes in chromatographic response were noted. For the robustness assessment, the initial detection wavelength (425 nm) was changed to detection wavelengths of 430 nm and 420 nm, and the differences in chromatographic response were recorded.

The standard deviation approach was used to evaluate the sensitivity of the greener HPLC methodology in terms of "limit of detection (LOD) and limit of quantitation (LOQ)" [57]. The standard deviation of the blank sample (without CCM) was computed following the injection of the blank sample into the HPLC apparatus three times ($n = 3$). After that, established techniques that have been documented in the literature were used to calculate the LOD and LOQ values for CCM [57,58].

The solution stability of CCM in stock solution and nanoemulsion was performed at the MQC level (20 µg/mL) at two distinct temperatures, namely, the bench temperature (25 $\pm$ 0.5 °C) and refrigeration temperature (4 $\pm$ 0.5 °C). These studies were carried out for the period of 72 h. The MQC concentration of CCM was freshly produced in a greener eluent system. The freshly prepared nanoemulsion was also diluted with the greener eluent system to obtain the MQC level of CCM. Both solutions were stored at 25 $\pm$ 0.5 °C and

4 ± 0.5 °C for 72 h and the decomposition of CCM was evaluated by measuring the rest of CCM after storage.

2.7. Forced Degradation/Selectivity Studies

Forced degradation studies under a variety of stress conditions, including acidic (HCl) stress, basic (NaOH) stress, oxidative (H_2O_2) stress, thermal stress, and photolytic stress conditions, were conducted in order to evaluate the selectivity and stability-indicating property of the greener HPLC methodology [56,60]. The degradation studies were performed in mild conditions, as recommended by ICH [57]. The 40 µg/mL of CCM was freshly prepared into the eluent system for acid and base-induced degradation. By mixing 4 mL of 1 M HCl and 4 mL of 1 M NaOH into an aliquot (1 mL) of this solution, acid and base hydrolysis were applied. For the determination of CCM in the presence of its acid- and base-degradation products, respectively, these mixtures were refluxed for 48 h at 60 °C before being evaluated using the greener HPLC approach [60].

The 40 µg/mL of CCM was freshly produced and introduced into the eluent system for oxidative degradation testing. This solution was oxidatively degraded by adding 4 mL of 30% H_2O_2 to an aliquot (1 mL) of it. For the detection of CCM in the presence of its oxidative-degradation products, this mixture was refluxed for 48 h at 60 °C before being evaluated using the greener HPLC approach [60].

The 40 µg/mL concentration of CCM (1.0 mL) was diluted with eluent system to produce a total volume of 5.0 mL. This solution was then subjected to a hot air oven for 48 h at 60 °C for thermal degradation tests. It was then assessed utilizing the ecofriendly HPLC technology for the detection of CCM in the presence of its thermal-degradation products [60].

For photolytic degradation investigations, a 1.0 mL aliquot of 40 µg/mL concentration was diluted with the eluent system to obtain the total volume of 5.0 mL. This solution was then subjected to a UV chamber at 254 nm for 48 h. Then, CCM was determined using the ecofriendly HPLC methodology while the photolytic-degradation products were present [60].

2.8. Greenness Measurement

The greenness profile for the ecofriendly HPLC methodology was determined using the "AGREE metric approach" [54]. The AGREE scores (0.0–1.0) were derived using the "AGREE: The Analytical Greenness Calculator (version 0.5, Gdansk University of Technology, Gdansk, Poland, 2020)".

2.9. Application of Greener HPLC Methodology in Determination of CCM in In-House Developed Nanoemulsion

In the lab, a CCM nanoemulsion was created and evaluated. One mL of an in-house developed nanoemulsion containing 20 mg/mL of CCM was appropriately diluted with the eluent system to produce 100 mL of stock solution in order to determine the CCM content. Following a suitable dilution with the eluent system and a 15-min sonication of this solution, the CCM content was determined using the ecofriendly HPLC approach. The potential for interference from components of the formulation that are nanoemulsions was also investigated.

2.10. Application of Greener HPLC Methodology in Determination of CCM in Curcuma Longa Extract

An amount of 1 mL of freshly prepared *Curcuma longa* extract was appropriately diluted with the eluent system to create 50 mL of stock solution in order to measure the CCM content. Following a suitable dilution with the eluent system and a 15-min sonication of this solution, the CCM content was determined using the ecofriendly HPLC approach.

2.11. Application of Greener HPLC Methodology in Determination of CCM in Commercial Tablets

The average mass of ten marketed tablets—each having 1000 mg of *C. longa* standardized extract—was determined. Ten tablets were crushed using a glass pestle and mortar to obtain fine powder. An amount of 100 mL of the eluent system was combined with a portion of the powder containing an average weight of the tablet. Then, 1 mL of this solution was added to 50 mL of the eluent system. To eliminate any insoluble excipient, the obtained solution of the commercial tablet was filtered using Whatman filter paper (No. 41) and sonicated at 25 °C for 25 min. Using the eco-friendly HPLC approach, the amount of CCM in commercial tablets was determined.

3. Results and Discussion

3.1. Analytical Method Development

Table 1 provides a summary of the measured chromatographic characteristics and the composition of distinct greener eluent systems. The use of methanol-water, ethanol-water, and acetone-water in different compositions during the analytical method development step resulted in a subpar chromatographic response of CCM, which was exhibiting higher As values (As > 2.0) with low N values (<2000). Furthermore, the use of methanol-ethanol, methanol-acetone, and methanol-ethyl acetate in distinct compositions caused CCM to have a poor chromatographic response, as well as increased As values (As > 1.20) and low N values (<3000). Further, the use of ethanol-acetone and ethyl acetate-acetone was also examined as the greener eluent systems. With high As values (As > 1.80) and low N values (<2500), the chromatographic response of CCM was once more subpar. However, a well-resolved and intact CCM chromatographic peak with good As values and greater N values was shown by the binary mixture of ethanol and ethyl acetate in distinct composition. The binary mixture of ethanol and ethyl acetate (83:17 v/v) gave the best chromatographic response and reliable retention time (R_t), as well as As and N values, among the various ethanol and ethyl acetate mixes examined (Figure 2). As a consequence, the binary combination of ethanol and ethyl acetate (83:17 v/v) was selected as the final greener eluent system for measuring CCM with an appropriate As (1.09) and N (5081), quick analysis (R_t = 3.57 min), and a good analysis period (5 min). The finally used solvents, such as ethanol and ethyl acetate, are non-toxic and environmentally safe [61,62]. Both solvents have been extensively studied as green solvents in the literature [58–62]. As a result, these solvents were used in this study to create a greener HPLC approach to determine CCM.

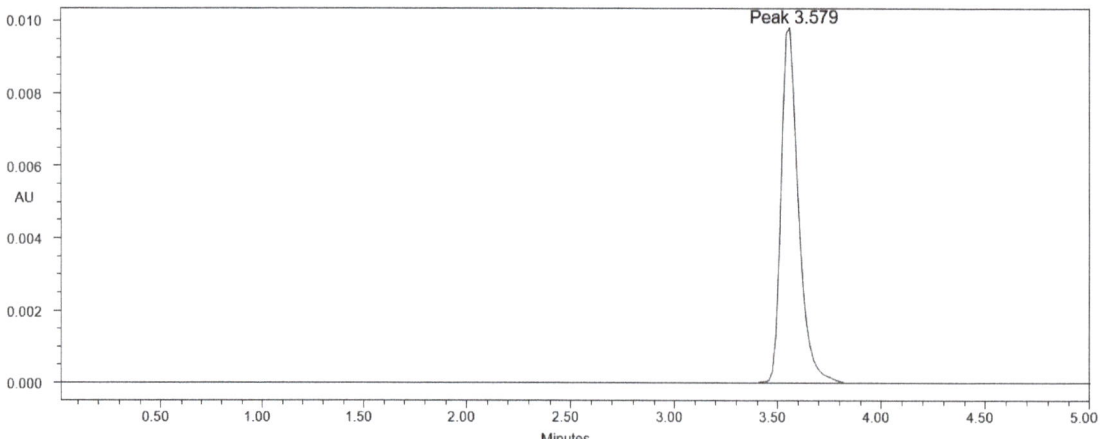

Figure 2. A greener high-performance liquid chromatography (HPLC) chromatogram of CCM in solution derived using ethanol: ethyl acetate (83:17 v/v) greener eluent system.

Table 1. The optimization of greener eluent systems and measured analytical responses for standard curcumin (CCM) (mean ± SD, n = 3).

Greener Eluent System	As	N	R_t
Methanol:water (50:50 v/v)	2.45 ± 0.36	1354 ± 2.31	5.26 ± 0.25
Methanol:water (83:17 v/v)	2.37 ± 0.31	1478 ± 2.38	5.04 ± 0.22
Ethanol:water (50:50 v/v)	2.14 ± 0.23	1741 ± 2.58	4.71 ± 0.18
Ethanol:water (83:17 v/v)	2.06 ± 0.19	1892 ± 2.62	4.42 ± 0.16
Acetone:water (50:50 v/v)	2.56 ± 0.41	1141 ± 2.01	6.16 ± 0.41
Acetone:water (83:17 v/v)	2.49 ± 0.38	1262 ± 2.13	6.08 ± 0.36
Methanol:ethanol (50:50 v/v)	1.78 ± 0.10	1924 ± 2.68	4.38 ± 0.20
Methanol:ethanol (83:17 v/v)	1.82 ± 0.12	1812 ± 2.59	4.41 ± 0.25
Methanol:acetone (50:50 v/v)	2.38 ± 0.28	1312 ± 2.21	6.74 ± 0.45
Methanol:acetone (83:17 v/v)	2.28 ± 0.24	1451 ± 2.30	6.69 ± 0.40
Methanol:ethyl acetate (50:50 v/v)	1.24 ± 0.08	2914 ± 3.27	4.16 ± 0.08
Methanol:ethyl acetate (83:17 v/v)	1.23 ± 0.07	2989 ± 3.38	4.10 ± 0.06
Ethanol:ethyl acetate (50:50 v/v)	1.13 ± 0.04	4125 ± 3.87	3.84 ± 0.03
Ethanol:ethyl acetate (83:17 v/v)	1.09 ± 0.02	5081 ± 5.48	3.57 ± 0.03
Ethanol:acetone (50:50 v/v)	2.07 ± 0.05	2122 ± 2.12	5.22 ± 0.08
Ethanol:acetone (83:17 v/v)	1.84 ± 0.06	2245 ± 2.31	5.11 ± 0.07
Ethyl acetate: acetone (50:50 v/v)	2.15 ± 0.09	1689 ± 2.27	6.41 ± 0.10
Ethyl acetate: acetone (83:17 v/v)	1.98 ± 0.08	1774 ± 2.18	6.02 ± 0.09

As: tailing factor; N: number of theoretical plates; R_t: retention time.

3.2. Validation Parameters

Numerous validation parameters for the greener HPLC methodology were assessed following ICH-Q2-R1 procedures [57]. The linearity graphs were produced using freshly prepared CCM samples (1–100 µg/mL). The outcomes of a linear regression analysis of the CCM calibration curve are summarized in Table 2. The linear calibration curve for CCM was between 1 and 100 µg/mL. According to estimates, the calibration curve's determination coefficient (R^2) and regression coefficient (R) values are 0.9983 and 0.9991, respectively, suggesting a good relation between CCM concentrations vs. the measured response. These data demonstrated the efficiency of the greener HPLC methodology for determining CCM.

Table 2. Linear regression data for the calibration curve of CCM for the greener high-performance liquid chromatography (HPLC) methodology (mean ± SD, n = 3).

Parameters	Values
Linearity range (µg/mL)	1–100
Regression equation	y = 70,341x − 59,690
R^2	0.9983
R	0.9991
SE of slope	58.96
SE of intercept	49.89
95% CI of slope	70,087–70,594
95% CI of intercept	59,475–59,904
LOD (µg/mL)	0.39 ± 0.03
LOQ (µg/mL)	1.17 ± 0.09

R^2: determination coefficient; R: regression coefficient; SE: standard error; CI: confidence interval; LOD: limit of detection; LOQ: limit of quantification.

The system suitability parameters for the greener HPLC methodology were estimated using the Rs, α, As, k, and N and results are summarized in Table 3. The greener HPLC methodology's values for Rs, peak symmetry, α, As, k, and N were found to be satisfactory and trustworthy for determining CCM.

Table 3. Optimized chromatographic peak parameters to determine CCM for the greener HPLC methodology (mean ± SD, n = 3).

Drug	Rs	α	As	k	N
PTT	2.41 ± 0.17	1.561 ± 0.09	1.09 ± 0.02	2.76 ± 0.15	5081 ± 5.48

Rs: resolution; α: selectivity; As: tailing factor; k: capacity factor; N: number of theoretical plates.

The percent recovery at three distinct QC levels was used to evaluate the intra-day and inter-day accuracy of the greener HPLC methodology. The results are summarized in Table 4. At three distinct QC levels, the intra-day and inter-day percent recoveries of CCM were determined to be 99.84–101.85 and 98.90–101.44 percent, respectively. High percent recoveries for the greener HPLC methodology for determining CCM point to its accuracy.

Table 4. Intra-day and inter-day accuracy data of CCM for the greener HPLC methodology (mean ± SD; n = 3).

Conc. (µg/mL)	Intra-Day Accuracy			Inter-Day Accuracy		
	Conc. Found (µg/mL) ± SD	Recovery (%)	CV (%)	Conc. Found (µg/mL) ± SD	Recovery (%)	CV (%)
15	15.24 ± 0.17	101.60	1.15	14.91 ± 0.18	99.40	1.20
20	20.37 ± 0.20	101.85	0.98	19.78 ± 0.21	98.90	1.06
25	24.96 ± 0.23	99.84	0.92	25.36 ± 0.25	101.44	0.98

The results of the intra-day and inter-day precisions are included in Table 5 and are expressed in %CV. For CCM, the intraday precision percent CVs were observed to range from 0.86 to 0.94%. Contrarily, the %CVs for inter-day precision ranged between 0.96 and 1.17%. Low %CVs in the greener HPLC methodology for determining CCM indicated its precision.

Table 5. Intra-day and inter-day precision data of CCM for the greener HPLC approach (mean ± SD; n = 3).

Conc. (µg/mL)	Intra-Day Precision			Inter-Day Precision		
	Conc. Found (µg/mL) ± SD	SE	CV (%)	Conc. Found (µg/mL) ± SD	SE	CV (%)
15	14.88 ± 0.14	0.08	0.94	15.21 ± 0.17	0.09	1.17
20	19.82 ± 0.18	0.10	0.90	20.28 ± 0.21	0.12	1.03
25	25.41 ± 0.22	0.12	0.86	24.86 ± 0.24	0.13	0.96

Table 6 summarizes the outcomes of the robustness evaluation for the MQC level of CCM. When evaluating robustness by altering the composition of the eluent system, the %CV and R_t were found to be 1.02–1.28% and 3.55–3.59 min, respectively. The %CV and R_t were calculated to be 1.16–1.18% and 3.23–3.81 min, respectively, in the scenario of a robustness assessment when the flow speed was altered. The %CV and R_t were found to be 1.14–1.22% and 3.54–3.60 min, respectively, in the scenario of the robustness assessment by altering the detecting wavelength. Low CVs and minimal R_t value variation in the greener HPLC methodology for detecting CCM point to its robustness.

The results of analyzing the sensitivity of the ecofriendly HPLC methodology in terms of "LOD and LOQ" are presented in Table 2. According to calculations, the "LOD and LOQ" using the ecofriendly HPLC approach are 0.39 ± 0.03 µg/mL and 1.17 ± 0.09 µg/mL, respectively. These results indicated that the ecofriendly HPLC technology would be sensitive enough to identify and measure CCM in a wide range of concentrations.

The stability of CCM in stock solution and nanoemulsion formulation at two distinct temperatures was also determined. The results of stability determination at two distinct temperatures are presented in Table 7. The CCM degradation was determined by measuring the rest of CCM concentration after storage. The CCM decomposition was very low when stored for 72 h at 25 ± 0.5 °C and at 4 ± 0.5 °C when the peak areas of the stored CCM solution and nanoemulsion were compared to those derived using a freshly prepared

CCM solution and CCM nanoemulsion. The precision of the CCM stock solution and nanoemulsion in terms of %CV was measured to be 0.71–0.80% and 0.74–0.79%, respectively, at two distinct temperatures. In addition, the % recovery of the CCM stock solution and nanoemulsion was determined to be 98.45–99.85% and 100.70–100.85%, respectively, at two distinct temperatures. CCM was found to be sufficiently stable in stock solution and nanoemulsion formulation at 25 and 4 °C based on these results.

Table 6. Robustness data of CCM at MQC (20 µg/mL) for the greener HPLC methodology (mean ± SD; n = 3).

Parameters	Conc. Found (µg/mL) ± SD	CV (%)	R_t ± SD	CV (%)
Greener eluent system (% v/v)				
(85:15)	18.61 ± 0.19	1.02	3.55 ± 0.03	0.84
(81:19)	21.08 ± 0.27	1.28	3.59 ± 0.02	0.55
Flow speed (mL/min)				
(1.15)	21.06 ± 0.25	1.18	3.23 ± 0.02	0.61
(0.85)	18.96 ± 0.22	1.16	3.81 ± 0.04	1.04
Measurement wavelength (nm)				
420	19.61 ± 0.24	1.22	3.54 ± 0.03	0.84
430	21.04 ± 0.24	1.14	3.60 ± 0.02	0.55

Table 7. Stability data of CCM in stock solution and nanoemulsion formulation at MCQ level at two different temperatures (mean ± SD; n = 3).

Sample Matrices	Nominal Conc. (µg/mL)	Conc. Found (µg/mL) ± SD	Precision (% CV)	Recovery (%)
		Stock solution		
Refrigeration (4 °C)	20	19.97 ± 0.16	0.80	99.85
Bench top (25 °C)	20	19.69 ± 0.14	0.71	98.45
		Nanoemulsion		
Refrigeration (4 °C)	20	20.14 ± 0.15	0.74	100.70
Bench top (25 °C)	20	20.17 ± 0.16	0.79	100.85

3.3. Forced Degradation and Selectivity Studies

By subjecting the 40 µg/mL concentration of CCM to various stress conditions, the selectivity and stability-indicating properties for the greener HPLC methodology were assessed. Figure 3 and Table 8 provide an overview of the outcomes of selectivity under various stress circumstances using the ecofriendly HPLC approach.

Table 8. Results of forced-degradation studies of CCM at 40 µg/mL concentration under various stress tests for the greener HPLC assay (mean ± SD; n = 3).

Stress Condition	CCM R_t (min)	CCM Remaining (µg/mL)	CCM Recovered (%)
1 M HCl	3.53	35.14	87.85 ± 1.91
1 M NaOH	3.56	33.61	84.02 ± 1.87
30% H_2O_2	3.57	38.74	96.85 ± 2.24
Thermal	3.57	39.97	99.92 ± 2.31
Photolytic	3.57	18.24	45.60 ± 1.71

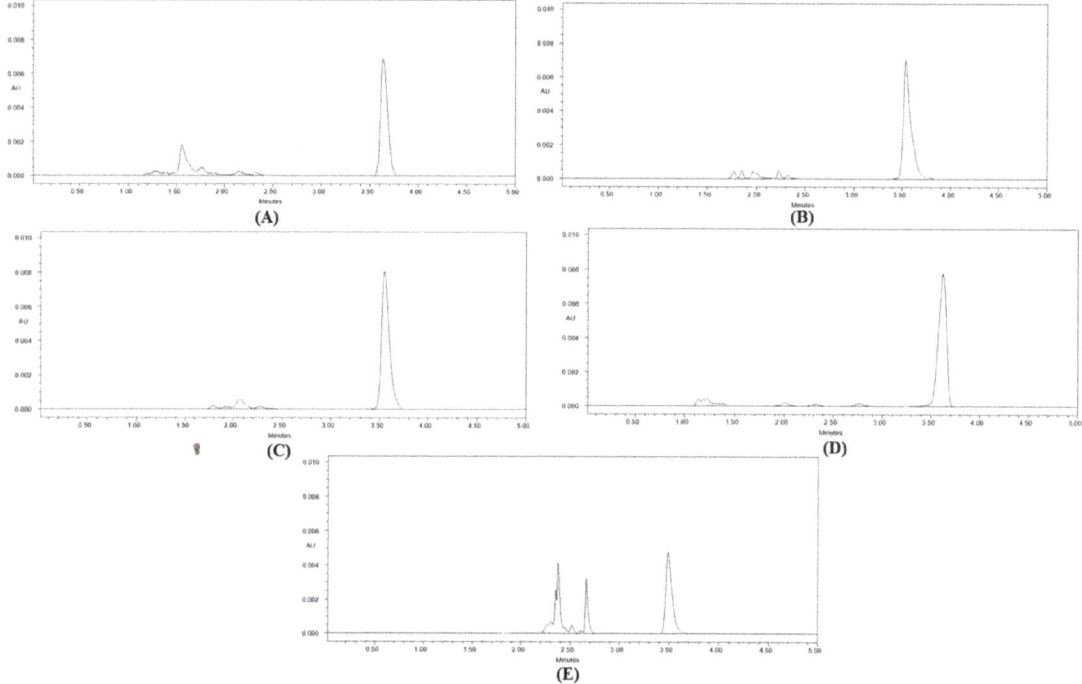

Figure 3. Greener HPLC chromatograms of CCM recorded under (**A**) acid-induced degradation, (**B**) base-induced degradation, (**C**) oxidative degradation, (**D**) thermal degradation, and (**E**) photolytic degradation of CCM.

The forced degradation investigations' chromatograms showed well-separated CCM peaks, along with a few other peaks of degradation products (Figure 3). In total, 87.85% of CCM was preserved under acid stress degradations, while 13.15 percent was degraded (Table 8). As a result, it was discovered that CCM was adequately stable under acidic degradations. The R_t value for CCM breakdown under acid stress was somewhat off (R_t = 3.53 min) (Figure 3A). In total, 84.02% of CCM was still present at base-stress degradations, while 15.98% was degraded (Table 8). As a result, it was discovered that CCM was adequately stable under alkali degradations. Additionally, a small shift was made to the R_t value of CCM during base-stress degradation (R_t = 3.56 min) (Figure 3B). Only 3.15% of CCM was found to be destroyed under oxidative-stress degradations, leaving 96.85% of the original material intact. As a result, it was discovered that CCM was adequately stable under oxidative stress degradation. The R_t value of CCM during oxidative stress degradation was not altered (R_t = 3.57 min) (Figure 3C). Under thermal degradation, 99.92% of CCM remained and only a negligible amount (0.08%) was degraded. Hence, CCM was found to be highly stable under thermal degradation. The R_t value of CCM under thermal degradation was not shifted (R_t = 3.57 min) (Figure 3D). Under photolytic degradation, 45.60% of CCM remained and 54.40% was degraded. As a result, CCM was found to be highly unstable under photolytic degradation. The R_t value of CCM under photolytic degradation was not shifted (R_t = 3.57 min) (Figure 3E). Overall, the maximum degradation of CCM was found under photolytic degradation. The degradation patterns of CCM were found to be identical with those reported previously in the literature [23]. Since the greener HPLC methodology was able to detect CCM in the presence of its degradation products, it can be considered as a stability-indicating one. Overall, these results indicated the selectivity and stability-indicating characteristics of the greener HPLC methodology.

3.4. Greenness Measurement

A number of analytical techniques are used to assess the greenness characteristics of analytical procedures [52–56]. Only the AGREE approach takes into account all twelve GAC principles while evaluating the analytical approaches' greenness [54]. As a result, the new HPLC methodology's greenness properties were identified utilizing the AGREE approach. The predicted overall AGREE score employing the twelve distinct GAC principles is summarized in Figure 4. Figure 5 includes the AGREE report sheet and AGREE score for each GAC concept. The different AGREE scores for each criteria of GAC were assigned by the AGREE calculator. The assigned scores ranged from 0.0 to 1.0. The established HPLC methodology's overall AGREE score was 0.81, indicating that it possesses exceptional greenness properties for the measurement of CCM. The AGREE score for the reported analytical methods of CCM measurement has not been determined in the literature. However, the AGREE score for the HPLC assay of some other drugs has been reported recently [58,59]. The AGREE score for the HPLC assay of the emtricitabine measurement has been reported to be 0.72 using the AGREE calculator [58]. Similarly, the AGREE score for the HPLC assay of the doxorubicin measurement has been reported to be 0.79 using the AGREE calculator [59]. The recorded AGREE score for the HPLC assay of the CCM measurement was better than those reported for the measurement of emtricitabine and doxorubicin [58,59]. Overall, the greenness profile of the current HPLC assay was outstanding.

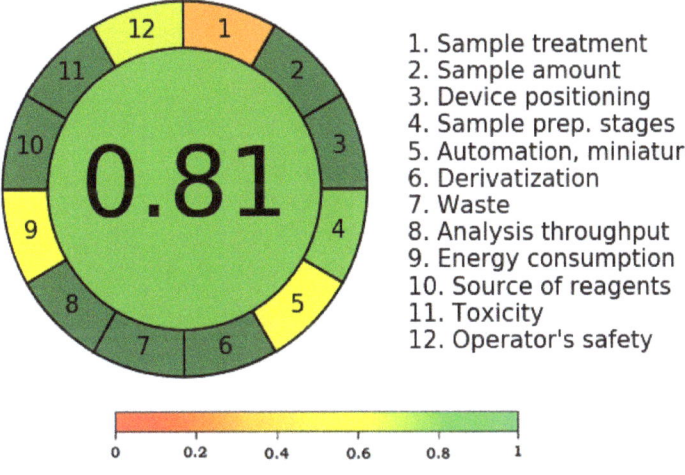

Figure 4. Analytical GREEnness (AGREE) score for the established HPLC assay of CCM determination predicted using AGREE calculator.

3.5. Determination of CCM in In-House Developed Nanoemulsion, Curcuma Longa Extract, and Commercial Tablets

The stability-indicating greener HPLC methodology for CCM determination was shown to be efficient, rapid, and sensitive. This technique was therefore applied to ascertain CCM in an in-house developed nanoemulsion, *Curcuma longa* extract, and commercial tablets. The CCM percentage assay was 101.24 ± 0.72% in the in-house developed nanoemulsion. The amount of CCM was determined to be 81.15 ± 0.64% in *Curcuma longa* extract. The amount of CCM in commercial tablets (containing 1000 mg of standardized *C. longa* extract) was found to be 78.41 ± 0.58%. The assay of CCM in ordinary emulsion formulation using the HPLC method has been reported to be 99.45% [23], which was similar to the present method of CCM assay in nanoemulsion formulation. The amount of CCM in six different commercial *C. longa* extracts using the HPLC method has been reported in the range of 69.82 to 86.79% [26]. The amount of CCM in *C. longa* extract using the present HPLC method was superior to most of the reported extracts [26]. Using another

HPLC method, the amount of CCM in *C. longa* extract has been reported as 30.76 mg/g (equivalent to 3.06%) [27], which was inferior to the present HPLC method. These findings indicated that the greener HPLC methodology would work well for figuring out CCM in its laboratory developed formulations, commercially available products, and distinct plant extracts.

Analytical Greenness report sheet

Criteria	Score	Weight
1. Direct analytical techniques should be applied to avoid sample treatment.	0.3	2
2. Minimal sample size and minimal number of samples are goals.	1.0	2
3. If possible, measurements should be performed in situ.	1.0	2
4. Integration of analytical processes and operations saves energy and reduces the use of reagents.	0.8	2
5. Automated and miniaturized methods should be selected.	0.5	2
6. Derivatization should be avoided.	1.0	2
7. Generation of a large volume of analytical waste should be avoided, and proper management of analytical waste should be provided.	1.0	2
8. Multi-analyte or multi-parameter methods are preferred versus methods using one analyte at a time.	1.0	2
9. The use of energy should be minimized.	0.5	2
10. Reagents obtained from renewable sources should be preferred.	1.0	2
11. Toxic reagents should be eliminated or replaced.	1.0	2
12. Operator's safety should be increased.	0.6	2

Figure 5. AGREE scale sheet presenting AGREE scale for 12 distinct components of GAC for the established HPLC methodology of CCM determination recorded using AGREE calculator.

4. Conclusions

The CCM in an in-house developed nanoemulsion, *Curcuma longa* extract, and commercial tablets has been identified and verified using a quick, sensitive, stability-indicating, and greener HPLC approach. The stability-indicating greener HPLC methodology was validated according to ICH-Q2-R1 procedures. For detecting CCM, the greener HPLC approach is more efficient, accurate, precise, stability-indicating, robust, sensitive, and selective. The greener HPLC methodology was found to be suitable for the determination of CCM in the in-house developed nanoemulsion, *Curcuma longa* extract, and commercial tablets. The AGREE evaluation suggested outstanding greenness characteristics of the established HPLC methodology. Because of its selectivity and stability-indicating properties, the greener HPLC method was able to identify CCM, even in the presence of its degradation products. Based on these findings, it is possible to effectively determine CCM in a variety of sample matrices using the stability-indicating greener HPLC methodology.

In future, further studies can be carried out to determine CCM in the complex matrices of biological samples and CCM pharmacokinetic evaluation.

Author Contributions: Conceptualization, S.A. and F.S.; methodology, N.H., M.M.G. and S.M.B.A.; software, N.H., S.A.A. and F.S.; validation, S.A. and P.A.; formal analysis, S.A.A. and P.A.; investigation, F.S. and N.H.; resources, S.A.; data curation, P.A.; writing—original draft preparation, F.S.; writing—review and editing, N.H., S.A. and P.A.; visualization, S.A.; supervision, F.S. and S.A.; project administration, F.S. and S.A.; funding acquisition, S.A. and M.M.G. All authors have read and agreed to the published version of the manuscript.

Funding: This study is supported via funding from Prince Sattam Bin Abdulaziz University project number (PSAU/2023/R/1444). The APC was funded by PSAU.

Institutional Review Board Statement: Not applicable.

Informed Consent Statement: Not applicable.

Data Availability Statement: Not applicable.

Acknowledgments: The authors are grateful to Prince Sattam Bin Abdulaziz University for supporting this work via project number (PSAU/2023/R/1444). The authors are also grateful to AlMaarefa University for their generous support.

Conflicts of Interest: The authors declare no conflict of interest.

References

1. El-Hack, M.E.A.; El-Saadony, M.T.; Swelum, A.A.; Arif, M.; Ghanima, M.M.A.; Shukry, M.; Noreldin, A.; Taha, A.E.; El-Tarabily, K.A. Curcumin, the active substance of turmeric: Its effects on health and ways to improve its bioavailability. *J. Sci. Food Agric.* **2021**, *101*, 5747–5762. [CrossRef] [PubMed]
2. Reeta, V.; Kalia, S. Turmeric: A review of its effects on human health. *J. Med. Plant Stud.* **2022**, *10*, 61–63.
3. Fabianowska-Majewska, K.; Kaufman-Szymczyk, A.; Szymanska-Kolba, A.; Jakubik, J.; Majeski, G. Curcumin from turmeric rhizome: A potential modulator of DNA methylation machinery in breast cancer inhibition. *Nutrients* **2021**, *13*, 332. [CrossRef] [PubMed]
4. Jayaprakasha, G.K.; Rao, L.J.M.; Sakariah, K.K. Improved HPLC method for the determination of curcumin, demethoxycurcumin, and bisdemethoxycurcumin. *J. Agric. Food Chem.* **2002**, *50*, 3668–3672. [CrossRef] [PubMed]
5. Wang, Y.; Wang, Y.; Cai, N.; Xu, T.; He, F. Anti-inflammatory effects of curcumin in acute lung injury: In vivo and in vitro experimental model studies. *Int. Immunopharmacol.* **2021**, *96*, E107600. [CrossRef]
6. Trigo-Gutierrez, J.K.; Vega-Chacon, Y.; Soares, A.B.; de Oliveira Mima, E.G. Antimicrobial activity inflammatory of curcumin in nanoformulations: A comprehensive review. *Int. J. Mol. Sci.* **2021**, *22*, 7130. [CrossRef]
7. Negahdari, R.; Ghavimi, M.A.; Barzegar, A.; Memar, M.Y.; Balazadeh, L.; Bohlouli, S.; Sharifi, S.; Dizaj, S.M. Antibacterial effects of nanocurcumin inside the implant fixture: An in vitro study. *Clin. Exp. Dental Res.* **2021**, *7*, 163–169. [CrossRef]
8. Ardebili, A.; Pouriayevali, M.H.; Aleshikh, S.; Zahani, M.; Ajorloo, M.; Izanloo, A.; Siyadatpanah, A.; Nikoo, H.R.; Wilairatana, P.; Coutinho, H.D.M. Antiviral therapeutic potential of curcumin: An update. *Molecules* **2021**, *26*, 6994. [CrossRef]
9. Urosevic, M.; Nikolic, L.; Gajic, I.; Nikolic, V.; Dinic, A.; Miljkovic, V. Curcumin: Biological activities and modern pharmaceuticals. *Antibiotics* **2022**, *11*, 135. [CrossRef]
10. Eremina, N.V.; Zhanataev, A.K.; Durnev, A.D. Induced cell death as a possible pathway of antimutagenic action. *Bull. Exp. Biol. Med.* **2021**, *171*, 1–14. [CrossRef]
11. Morshidi, K.; Borran, S.; Ebrahimi, M.S.; Khooy, M.J.M.; Seyedi, Z.S.; Amiri, A.; Abbasi-Kolli, M.; Fallah, M.; Khan, H.; Sahebkar, A.; et al. Therapeutic effect of curcumin in gastrointestinal cancers: A comprehensive review. *Phytother. Res.* **2021**, *35*, 3834–3897. [CrossRef] [PubMed]
12. Zhang, X.; Yang, X.; Lin, J.; Song, F.; Shao, Y. Low curcumin concentration enhances the anticancer effects of 5-fluorouracil against colorectal cancer. *Phytomedicine* **2021**, *85*, E153547. [CrossRef] [PubMed]
13. Arena, A.; Romeo, M.A.; Benedetti, R.; Masuelli, L.; Bei, R.; Montani, M.S.G.; Cirone, M. New insights into curcumin- resveratrol-mediated anticancer effects. *Pharmaceuticals* **2021**, *14*, 1068. [CrossRef]
14. Unsal, Y.E.; Tuzen, M.; Soylak, M. Ultrasound-assisted ionic liquid-dispersive liquid–liquid of curcumin in food samples microextraction and its spectrophotometric determination. *J. AOAC Int.* **2019**, *102*, 217–221. [CrossRef] [PubMed]
15. Altunay, N.; Unal, Y.; Elik, A. Towards green analysis of curcumin from tea, honey and spices: Extraction by deep eutectic solvent assisted emulsification liquid-liquid microextraction method based on response surface design. *Food Addit. Cont. Part A.* **2020**, *37*, 869–881. [CrossRef] [PubMed]
16. Gurmen, K.; Sahin, U.; Yilmaz, E.; Soylak, M.; Sahan, S. Determination of curcumin in food with homogenous liquid-phase microextraction preconcentration and spectrophotometric determination. *Anal. Lett.* **2023**, *56*, 807–815. [CrossRef]

17. Karimi, M.; Mashreghi, M.; Saremi, S.S.; Jaafari, M.R. Spectrofluorometric method development and validation for the determination of curcumin in nanoliposomes and plasma. *J. Fluoresc.* **2020**, *30*, 1113–1119. [CrossRef]
18. Yang, R.; Mu, W.-Y.; Chen, Q.-Y. Urazole-Au nanocluster as a novel fluorescence probe for curcumin determination and mitochondria imaging. *Food Anal. Methods* **2019**, *12*, 1805–1812. [CrossRef]
19. Yu, C.; Zhuang, Q.; Cui, H.; Li, Y.; Ding, Y.; Lin, J.; Duan, Y. A fluorescent "turn-off" probe for the determination of curcumin using upconvert luminescent carbon dots. *J. Fluoresc.* **2020**, *30*, 1469–1476. [CrossRef]
20. Naidu, M.M.; Shyamala, B.N.; Manjunatha, J.R.; Sulochanamma, G.; Srinivas, P. Simple HPLC method for resolution of curcuminoids with antioxidant potential. *J. Food Sci.* **2009**, *74*, C312–C318. [CrossRef]
21. Jangle, R.D.; Thorat, B.N. Reversed-phase high-performance liquid chromatography method for analysis of curcuminoids and curcuminoid-loaded liposome formulation. *Indian J. Pharm. Sci.* **2013**, *75*, 60–66. [PubMed]
22. Ali, I.; Haque, A.; Saleem, K. Separation and identification of curcuminoids in turmeric powder by HPLC using phenyl column. *Anal. Methods* **2014**, *6*, 2526–2536. [CrossRef]
23. Syed, H.K.; Bin Liew, K.; Loh, G.O.K.; Peh, K.K. Stability indicating HPLC–UV method for detection of curcumin in *Curcuma longa* extract and emulsion formulation. *Food Chem.* **2015**, *170*, 321–326. [CrossRef]
24. Hwang, K.-W.; Son, D.; Jo, H.-W.; Kim, C.H.; Seong, K.C.; Moon, J.-K. Levels of curcuminoid and essential oil compositions in turmerics (*Curcuma longa* L.) grown in Korea. *Appl. Biol. Chem.* **2016**, *59*, 209–215. [CrossRef]
25. Osorio-Tobon, J.F.; Carvalho, P.I.N.; Barbero, G.F.; Nogueira, G.C.; Rostango, M.A.; Meireles, M.A.D.A. Fast analysis of curcuminoids from turmeric (*Curcuma longa* L.) by high-performance liquid chromatography using a fused-core column. *Food Chem.* **2016**, *200*, 167–174. [CrossRef]
26. Peram, M.R.; Jalalpure, S.S.; Joshi, S.A.; Palkar, M.B.; Diwan, P.V. Single robust RP-HPLC analytical method for quantification of curcuminoids in commercial turmeric products, Ayurvedic medicines, and nanovesicular systems. *J. Liq. Chromatogr. Relat. Technol.* **2017**, *40*, 487–498. [CrossRef]
27. Erpina, E.; Rafi, M.; Darusman, L.K.; Vitasari, A.; Putra, B.R.; Rohaeti, E. Simultaneous quantification of curcuminoids and xanthorrhizol in *Curcuma xanthorrhiza* by high-performance liquid chromatography. *J. Liq. Chromatogr. Relat. Technol.* **2017**, *40*, 635–639. [CrossRef]
28. Jiang, H.; Somogyi, A.; Jacobsen, N.E.; Timmermann, B.N.; Gang, D.R. Analysis of curcuminoids by positive and negative electrospray ionization and tandem mass spectrometry. *Rapid Commun. Mass Spectrom.* **2006**, *20*, 1001–1012. [CrossRef]
29. Jiang, H.; Timmermann, B.N.; Gang, D.R. Use of liquid chromatography–electrospray ionization tandem mass spectrometry to identify diarylheptanoids in turmeric (*Curcuma longa* L.) rhizome. *J. Chromatogr. A.* **2006**, *1111*, 21–31. [CrossRef]
30. Avula, B.; Wang, Y.-H.; Khan, I.A. Quantitative determination of curcuminoids from the roots of *Curcuma longa*, *Curcuma* species and dietary supplements using an UPLC-UV-MS method. *J. Chromatogr. Sep. Tech.* **2012**, *3*, E1000120. [CrossRef]
31. Cheng, J.; Weijun, K.; Yun, L.; Jiabo, W.; Haitao, W.; Qingmiao, L.; Xiaohe, X. Development and validation of UPLC method for quality control of *Curcuma longa* Linn.: Fast simultaneous quantitation of three curcuminoids. *J. Pharm. Biomed. Anal.* **2010**, *53*, 43–49. [CrossRef] [PubMed]
32. Ansari, M.J.; Ahmad, S.; Kohli, K.; Ali, J.; Khar, R.K. Stability-indicating HPTLC determination of curcumin in bulk drug and pharmaceutical formulations. *J. Pharm. Biomed. Anal.* **2005**, *39*, 132–138. [CrossRef] [PubMed]
33. Pathania, V.; Gupta, A.P.; Singh, B. Improved HPTLC method for determination of curcuminoids from *Curcuma longa*. *J. Liq. Chromatogr. Relat. Technol.* **2006**, *29*, 877–887. [CrossRef]
34. Green, C.E.; Hibbert, S.L.; Bailey-Shaw, Y.A.; Williams, L.A.D.; Mitchell, S.; Garraway, E. Extraction, processing, and storage effects on curcuminoids and oleoresin yields from *Curcuma longa* L. grown in Jamaica. *J. Agric. Food Chem.* **2008**, *56*, 3664–3670. [CrossRef]
35. Paramasivam, M.; Poi, R.; Banerjee, H.; Bandyopadhyay, A. High-performance thin layer chromatographic method for quantitative determination of curcuminoids in *Curcuma longa* germplasm. *Food Chem.* **2009**, *113*, 640–644. [CrossRef]
36. Taha, M.N.; Krawinkel, M.B.; Morlock, G.E. High-performance thin-layer chromatography linked with (bio)assays and mass spectrometry—A suited method for discovery and quantification of bioactive components? Exemplarily shown for turmeric and milk thistle extracts. *J. Chromatogr. A.* **2015**, *1394*, 137–147. [CrossRef]
37. Baghel, U.S.; Nagar, A.S.; Pannu, M.S.; Singh, D.; Yadav, R. HPLC and HPTLC methods for simultaneous estimation of quercetin and curcumin in polyherbal formulation. *Indian J. Pharm. Sci.* **2017**, *79*, 197–203. [CrossRef]
38. Chen, W.; Fan-Havard, P.; Yee, L.D.; Cao, Y.; Stoner, G.D.; Chan, K.K.; Liu, Z. A liquid chromatography–tandem mass spectrometric method for quantification of curcumin-O-glucuronide and curcumin in human plasma. *J. Chromatogr. B.* **2012**, *900*, 89–93. [CrossRef]
39. Kunati, S.R.; Yang, S.; William, B.M.; Xu, Y. An LC–MS/MS method for simultaneous determination of curcumin, curcumin glucuronide and curcumin sulfate in a phase II clinical trial. *J. Pharm. Biomed. Anal.* **2018**, *156*, 189–198. [CrossRef]
40. Liu, Y.; Siard, M.; Adams, A.; Keowen, M.L.; Miller, T.K.; Garza, F., Jr.; Andrews, F.M.; Seeram, N.P. Simultaneous quantification of free curcuminoids and their metabolites in equine plasma by LC-ESI–MS/MS. *J. Pharm. Biomed. Anal.* **2018**, *154*, 31–39. [CrossRef]
41. Yu, W.; Wen, D.; Cai, D.; Zheng, J.; Gan, H.; Jiang, F.; Liu, X.; Lao, B.; Yu, W.; Guan, Y.; et al. Simultaneous determination of curcumin, tetrahydrocurcumin, quercetin, and paeoniflorin by UHPLC-MS/MS in rat plasma and its application to a pharmacokinetic study. *J. Pharm. Biomed. Anal.* **2019**, *172*, 58–66. [CrossRef] [PubMed]

42. Hua, Q.; Gao, L.; Raoa, S.-Q.; Yanga, Z.-Q.; Li, T.; Gong, X. Nitrogen and chlorine dual-doped carbon nanodots for determination of curcumin in food matrix via inner filter effect. *Food Chem.* **2019**, *280*, 195–202. [CrossRef] [PubMed]
43. Deng, P.; Wei, Y.; Li, W.; Shi, S.; Zhou, C.; Li, J.; Yao, L.; Ding, J.; He, Q. A novel platform based on MnO_2 nanoparticles and carboxylated multi-walled carbon nanotubes composite for accurate and rapid determination of curcumin in commercial food products. *J. Food Compos. Anal.* **2023**, *115*, E104940. [CrossRef]
44. Raril, C.; Manjunatha, J.G.; Tigari, G. Low-cost voltammetric sensor based on ananionic surfactant modified carbon nanocomposite material for the rapid determination of curcumin in natural food supplement. *Inst. Sci. Technol.* **2020**, *48*, 561–582. [CrossRef]
45. Burc, M.; Gungor, O.; Duran, S.T. Voltammetric determination of curcumin in spices using platinum electrode electrochemically modified with poly(vanillin-co-caffeic acid). *Anal. Bioanal. Electrochem.* **2020**, *12*, 625–643.
46. Pushpanjali, P.A.; Manjunatha, J.G.; Amrutha, B.M.; Hareesha, N. Development of carbon nanotube-based polymer-modified electrochemical sensor for the voltammetric study of Curcumin. *Mater. Res. Innov.* **2021**, *25*, 412–420. [CrossRef]
47. Tigari, G.; Manjunatha, J.G. Poly(glutamine) film-coated carbon nanotube paste electrode for the determination of curcumin with vanillin: An electroanalytical approach. *Mon. Chem.* **2020**, *151*, 1681–1688. [CrossRef]
48. Rahimnejad, M.; Zokhtareh, R.; Moghadamnia, A.A.; Asghary, M. An electrochemical sensor based on reduced grapheme oxide modified carbon paste electrode for curcumin determination in human blood serum. *Purtug. Electrochem. Acta* **2020**, *38*, 29–42. [CrossRef]
49. Thangavel, K.; Dhivya, K. Determination of curcumin, starch and moisture content in turmeric by Fourier transform near infrared spectroscopy (FT-NIR). *Eng. Agric. Environ. Food* **2019**, *12*, 264–269. [CrossRef]
50. Gong, X.; Wang, H.; Liu, Y.; Hu, Q.; Gao, Y.; Yang, Z.; Shuang, S.; Dong, C. A di-functional and label-free carbon-based chem-nanosensor for real-time monitoring of pH fluctuation and quantitative determining of curcumin. *Anal. Chim. Acta* **2019**, *1057*, 132–144. [CrossRef]
51. Lizonova, D.; Brejchova, A.; Kralova, E.; Hanus, J.; Stepanek, F. Solvatochromic shift enables intracellular fate determination of curcumin nanocrystals. *Partic. Partic. Sys. Character.* **2022**, *39*, E2200115. [CrossRef]
52. Abdelrahman, M.M.; Abdelwahab, N.S.; Hegazy, M.A.; Fares, M.Y.; El-Sayed, G.M. Determination of the abused intravenously administered madness drops (tropicamide) by liquid chromatography in rat plasma; an application to pharmacokinetic study and greenness profile assessment. *Microchem. J.* **2020**, *159*, E105582. [CrossRef]
53. Duan, X.; Liu, X.; Dong, Y.; Yang, J.; Zhang, J.; He, S.; Yang, F.; Wang, Z.; Dong, Y. A green HPLC method for determination of nine sulfonamides in milk and beef, and its greenness assessment with analytical eco-scale and greenness profile. *J. AOAC Int.* **2020**, *103*, 1181–1189. [CrossRef] [PubMed]
54. Pena-Pereira, F.; Wojnowski, W.; Tobiszewski, M. AGREE-Analytical GREEnness metric approach and software. *Anal. Chem.* **2020**, *92*, 10076–10082. [CrossRef]
55. Alam, P.; Salem-Bekhit, M.M.; Al-Joufi, F.A.; Alqarni, M.H.; Shakeel, F. Quantitative analysis of cabozantinib in pharmaceutical dosage forms using green RP-HPTLC and green NP-HPTLC methods: A comparative evaluation. *Sus. Chem. Pharm.* **2021**, *21*, E100413. [CrossRef]
56. Foudah, A.I.; Shakeel, F.; Alqarni, M.H.; Alam, P. A rapid and sensitive stability-indicating green RP-HPTLC method for the quantitation of flibanserin compared to green NP-HPTLC method: Validation studies and greenness assessment. *Microchem. J.* **2021**, *164*, E105960. [CrossRef]
57. ICH. International Conference on Harmonization (ICH), Q2 (R1): Validation of Analytical Procedures—Text and Methodology. ICH: Geneva, Switzerland, 2005.
58. Haq, N.; Alshehri, S.; Alam, P.; Ghoneim, M.M.; Hasan, Z.; Shakeel, F. Green analytical chemistry approach for the determination of emtricitabine in human plasma, formulations, and solubility study samples. *Sus. Chem. Pharm.* **2022**, *26*, E100648. [CrossRef]
59. Haq, N.; Alanazi, F.K.; Samem-Bekhit, M.M.; Rabea, S.; Alam, P.; Alsarra, I.A.; Shakeel, F. Greenness estimation of chromatographic assay for the determination of anthracycline-based antitumor drug in bacterial ghost matrix of *Salmonella typhimurium*. *Sus. Chem. Pharm.* **2022**, *26*, E100642. [CrossRef]
60. Haq, N.; Iqbal, M.; Alanazi, F.K.; Alsarra, I.A.; Shakeel, F. Applying green analytical chemistry for rapid analysis of drugs: Adding health to pharmaceutical industry. *Arabian J. Chem.* **2017**, *10*, S777–S785. [CrossRef]
61. Hackl, K.; Kunz, W. Some aspects of green solvents. *Comp. Rend. Chim.* **2018**, *21*, 572–580. [CrossRef]
62. Welton, T. Solvents and sustainable chemistry. *Proc. Math Phys. Eng. Sci.* **2015**, *471*, E2015052. [CrossRef] [PubMed]

Disclaimer/Publisher's Note: The statements, opinions and data contained in all publications are solely those of the individual author(s) and contributor(s) and not of MDPI and/or the editor(s). MDPI and/or the editor(s) disclaim responsibility for any injury to people or property resulting from any ideas, methods, instructions or products referred to in the content.

Article

Analysis of Sugars in Honey Samples by Capillary Zone Electrophoresis Using Fluorescence Detection

Melinda Andrasi *, Gyongyi Gyemant, Zsofi Sajtos and Cynthia Nagy

Department of Inorganic and Analytical Chemistry, University of Debrecen, Egyetem ter 1., H-4032 Debrecen, Hungary
* Correspondence: andrasi.melinda@science.unideb.hu

Abstract: The applicability of capillary electrophoresis (CE) with light-emitting diode-induced fluorescence detection (LEDIF) for the separation of sugars in honey samples was studied. An amount of 25 mM ammonium acetate (pH 4.5) with 0.3% polyethylene oxide (PEO) was found to be optimal for the efficient separation of carbohydrates. 8-aminopyrene-1,3,6-trisulfonic acid (APTS) was used for the labeling of the carbohydrate standards and honey sugars for fluorescence detection. The optimized method was applied in the quantitative analysis of fructose and glucose by direct injection of honey samples. Apart from the labeling reaction, no other sample preparation was performed. The mean values of the fructose/glucose ratio for phacelia honey, acacia honey and honeydew honey were 0.86, 1.61 and 1.42, respectively. The proposed method provides high separation efficiency and sensitive detection within a short analysis time. Apart from the labeling reaction, it enables the injection of honeys without sample pretreatment. This is the first time that fluorescence detection has been applied for the CE analysis of sugars in honeys.

Keywords: capillary electrophoresis; fluorescence detection; honey; sugars

Citation: Andrasi, M.; Gyemant, G.; Sajtos, Z.; Nagy, C. Analysis of Sugars in Honey Samples by Capillary Zone Electrophoresis Using Fluorescence Detection. *Separations* **2023**, *10*, 150. https://doi.org/10.3390/separations10030150

Academic Editor: Faiyaz Shakeel

Received: 7 February 2023
Revised: 15 February 2023
Accepted: 19 February 2023
Published: 23 February 2023

Copyright: © 2023 by the authors. Licensee MDPI, Basel, Switzerland. This article is an open access article distributed under the terms and conditions of the Creative Commons Attribution (CC BY) license (https://creativecommons.org/licenses/by/4.0/).

1. Introduction

Honey is a natural, aqueous, supersaturated sugar substance produced by honeybees. It also contains other minor substances, such as minerals, enzymes, vitamins, organic acids, amino acids, flavonoids and phenolic acids. Honey is used as a nutritional product, but it can also exert several health-benefitting effects. The main components of honey are carbohydrates, representing around 85–97% of its weight. A significant part is made up of mainly fructose and glucose. Small amounts of other monosaccharides (galactose, mannose), disaccharides (maltose, sucrose, beta-trehalose) and oligosaccharides (melezitose, maltotriose, panose, erlose, isomaltotriose, theanderose) are present in honey [1,2].

The determination of sugars is a common approach for describing the quality of honey. The principle of classical chemical methods for the analysis of carbohydrates is based on the fact that reducing sugars (non-reducing carbohydrates can become reducing via hydrolyzation) react with other compounds to form precipitates or colored complexes, which can be determined by titration, gravimetric and spectrophotometric techniques [3,4]. The main disadvantages of the classical chemical methods are that carefully controlled, time consuming reaction parameters must be provided.

Separation techniques are the most powerful methods for the identification and quantification of carbohydrates, of which chromatographic methods, especially normal phase [5], anion-exchange [6], and hydrophilic interaction [7] high-performance liquid chromatography have been employed. A refractive index detector is commonly used in chromatographic analysis of sugars [8–10].

Capillary electrophoresis (CE) has emerged as an alternative tool for the analysis of carbohydrates. It provides high separation efficiency within a short analysis time, and low sample volume consumption. CE allows for the direct injection of real samples without any pretreatment. On the other hand, carbohydrate analysis by CE is quite challenging in terms of separation and detection. Most carbohydrates have no charge or absorbing chromophores in the UV-VIS regions. In recent years, different strategies have been developed to overcome these limitations [11–13]. The use of borate buffers with direct UV detection is the simplest way to analyze carbohydrates. Borate forms complexes with the vicinal hydroxyl groups of sugars, converting them into anions. This complex shows increased UV sensitivity at 191–195 nm [14]. The detection of carbohydrate-borate complexes can also be carried out using electrochemical methods (amperometric, contactless conductivity) [15]. Another possibility for making carbohydrates amenable to CE analysis is the use of strong alkaline background electrolytes (BGEs). Using a BGE with a pH above the pKa of the sugars ensures that their hydroxyl groups are dissociated; hence, sugars can migrate as anions in capillary zone electrophoresis (CZE) [16]. Different detection modes can be applied, such as indirect UV [17], fluorescence detection [18], electrochemical detection [19], and mass spectrometry (MS) [20]. The main merit of the indirect UV mode is that no time-consuming derivatization procedure is required; however, the limit of detection (LOD) is weaker than that of fluorescence detection. The laser-induced fluorescence (LIF) or LEDIF detection mode can ensure high sensitivity. The most frequent fluorescent labeling reaction is reductive amination [21]. There are several types of labeling reagents (8-aminopyrene-1,3,6-trisulfonic acid (APTS), 8-aminonaphthalene-1,3,6-trisulfonic acid (ANTS), aminonaphthalene-2-sulfonic acid (ANA), and 4-amino-5-hydroxynaphthalene-2,7-disulfonic acid (AHNS)) that can be used depending on the excitation and emission wavelengths [22]. APTS is a popular derivatizing agent because it can be excited with the commonly used argon-ion laser set at 488 nm and emitting at 520 nm. Furthermore, it adds three negative charges to the molecule, allowing for the possibility of CZE separation [23–25].

There are many examples of CE-based honey analysis in the literature, but very few of them deal with the determination of sugars. Indirect CE–UV was used by Rizelio et al. for the determination of fructose, glucose and sucrose in seven multifloral honey samples using 20 mM sorbic acid, 0.2 mM cetyltrimethylammonium bromide (CTAB) and 40 mM sodium hydroxide (NaOH) at pH 12.2. The detection limits for the three analytes were in the range of 0.022 to 0.029 g/L [26]. The drawbacks of indirect CE-UV are that the baseline is often not stable, and the sensitivity is worse compared to the fluorescence detection mode. A BGE consisting of 10 mM sodium benzoate and 1.5 mM CTAB, with a pH of 12.4, was applied for the simultaneous determination of fructose, glucose and sucrose in honey, using the indirect UV detection mode. The LOD for fructose, glucose and sucrose were 0.58 g/L, 0.67 g/L and 0.12 g/L, respectively [27]. The molar ratios of carbohydrates in 10 kinds of honey were determined by CZE after reductive, UV-active derivatization reactions with 1-phenyl-3-methyl-5-pyrazolone (PMP). Eleven PMP-labeled aldoses were separated in 200 mM borate-4% methanol at pH 11.0 [28]. A graphene–cobalt microsphere (CoMS) hybrid paste electrode was used for the detection of carbohydrates in three honey samples, for which the separation medium was 75 mM NaOH [29].

To the best of our knowledge, fluorescence detection has not yet been applied for the analysis of sugars in honeys. In this work, we applied APTS for the labeling of honey sugars. We optimized a CZE method for the separation of labeled sugars. The aim of this work was to demonstrate the applicability of CE for the determination of APTS-labeled sugars in honey samples.

2. Materials and Methods
2.1. Chemicals

All chemicals were of analytical grade. Ammonium acetate, hydrochloric acid (HCl), NaOH, PEO (average Mv~90,000 g/mol), 6-aminocaproic acid (EACA), hydroxypropyl methylcellulose (HPMC), APTS, sodium cyanoborohydride (NaBH$_3$CN), tetrahydrofuran (THF), acetic acid, fructose, mannose, glucose, galactose, maltose, arabinose, xylose, ribose, and galactose were obtained from Sigma Aldrich (St. Louis, MO, USA). Different types of honey samples (phacelia honey, acacia honey, honeydew honey) were a kind offer from

Hungarian producers. The ladder standard, malto-oligosaccharides composed of 1 to 7 glucose units, was made from β-cyclodextrin by acetolysis.

2.2. Instrumentation

CE separations were carried out using a 7100 CE System (Agilent, Waldbronn, Germany) coupled to UV and LEDIF (Zetalif Picometrics) detectors. Separations were performed using a fused-silica capillary (Polymicro, Phoenix, AZ, USA) of 65 cm, 50 cm × 50 μm i.d. The precondition procedure was 1 M HCl for 5 min, BGE for 5 min. Hydrodynamic sample injection (50 mbar × 2 s) was carried out at the anodic end of the fused silica capillary. For the electrophoretic separation, −30 kV was used. Fluorescence detection was performed by an LED-induced fluorescence detector; the excitation and emission wavelengths were 480 nm and 520 nm, respectively. The photomultiplier's high voltage was set to 700 V. For UV, on-capillary detection at a wavelength of 240 nm was chosen. Chemstation version B.04.02 software (Agilent) was used for operating the CE instrument and for processing the results.

2.3. Sample Preparation for Fluorescence Detection

The derivatization by APTS was carried out according to the recipe of Evangelista [30]; only minor modifications were made, mainly regarding the incubation and the amount of reagents. For the labeling reaction, 0.5 mg of the carbohydrate standard and 1 mg of the honey sample were measured and were dissolved in a mixture of 6 μL of 20 mM APTS (in 15% acetic acid) and 2 μL of 1 M $NaBH_3CN$ (in THF) solution. In the first step, a Schiff base was formed, which was reduced to a secondary amine by $NaBH_3CN$. The samples were homogenized using a vortex and then incubated for 1 h at 50 °C. The reaction was stopped by adding 92 μL of distilled water into the labeling reaction. The reaction mixtures were further diluted before direct sample injection.

3. Results and Discussion

3.1. CZE Separations of APTS-Labeled Carbohydrates

Fluorescence detection provides extremely sensitive detection for the analysis of sample injection plugs in the nanoliter range. As depicted in Figure 1, for the carbohydrate components, the LEDIF detection resulted in a high analytical response signal with very little baseline noise (Figure 1A), while detection at 240 nm yielded a small signal with high baseline noise (Figure 1B). The signal-to-noise ratio was much lower (S/N = 30) compared to fluorescence detection, where S/N was 81,000. Although UV detection is made possible by derivatization, it allows three orders of magnitude of lower sensitivity based on the calculated S/N values.

When labeling carbohydrates, derivatization should not simply be done for detectability; it is crucial to select a fluorophore that renders the labeled sugar component charged. The tagged sugar molecule gains three negative charges, since APTS has three sulfonate groups, enabling their migration in the electric field. Good resolution was achieved in a short migration window using a simple acetate buffer (pH 4.5) in the case of the analysis of the carbohydrate ladder containing oligomers of 1 to 7 glucose units (Figure 1). The electric field is applied under reversed polarity to drive anions toward the detection window. The separation relies on differences in charge-to-size ratio. All labeled sugar components possess a triple negative charge, but their molecular masses are different, so their electrophoretic mobilities differ, as well. The characteristic electrophoretic peak pattern is due to the method of sample production. The intensity decreases as the number of glucose units decreases. After the ring opening of β-CD, with 7 glucose units, an oligosaccharide composed of 7 glucose units was formed in the largest amount. During acidic decomposition, a decreasing amount of shorter oligosaccharides was formed, and glucose was present in high concentration in the sample due to the cleavage of glucose units (Figure 1B).

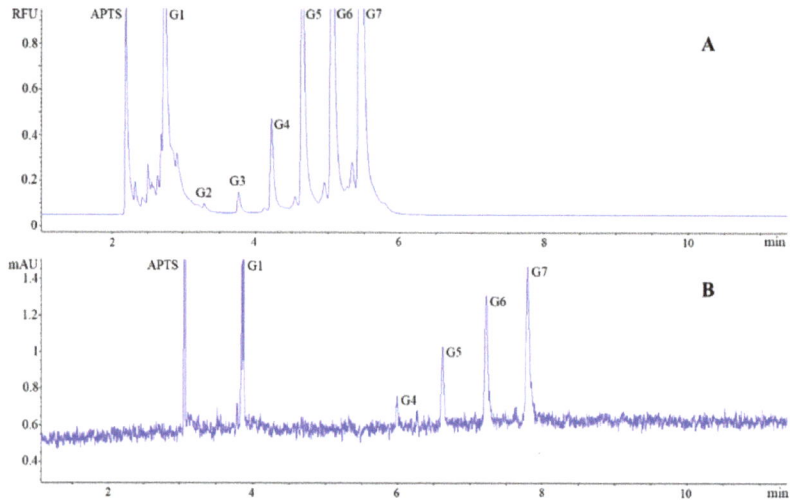

Figure 1. The CZE electropherograms of an APTS-labeled malto-oligosaccharide ladder standard using LEDIF (**A**) and UV (**B**) detection. The GX numbers indicate the numbers of glucose residues of the ladder. Conditions: fused silica capillary, l_{tot}: 50 cm × 50 μm id, l_{eff}: 30 cm (**A**), 42 cm (**B**), BGE: 25 mM ammonium acetate (pH 4.5), separation voltage: −30 kV, injection: 50 mbar × 2 s, preconditioning: 1 M HCl for 5 min, BGE for 5 min washing. LEDIF detection was performed at 480 nm excitation/520 nm emission, UV detection was performed at 240 nm.

A four-component carbohydrate mixture (glucose, fructose, mannose and maltose) most often found in honey was investigated using the same BGE that proved optimal in separating the members of the carbohydrate ladder (Figure 2A). The disaccharide maltose appeared as a peak well separated from the monosaccharides upon application of a simple ammonium acetate buffer, but no adequate resolution was obtained in the case of the three monosaccharides (Figure 2A). All three monosaccharides are hexoses ($C_6H_{12}O_6$), and their molar masses are the same (M = 180 g/mol); hence, there is no big difference in their electrophoretic mobility, which explains the merging of the peaks. The separation of fructose from mannose and glucose can be attributed to the fact that electrophoretic mobility is also determined by the shape and hydrodynamic radius of the particle. The use of simple, additive-free buffers does not always provide adequate resolution for sugars with the same number of carbon atoms [31].

In order to increase the selectivity, PEO (average Mv~90,000) was added into the BGE, as a result of which a suitable resolution was achieved for the three hexoses (Figure 2B). The effect of PEO on the enhancement of the separation is complex [32]. Due to its neutral polymer nature, it is connected to the inner surface of the fused silica capillary by secondary bonds, so it has a surface modification effect. The neutral coating suppresses the adsorption of the negatively charged components to protonated silanol groups. PEO also has a sieving effect by increasing the viscosity of the BGE. The increase in migration times is due to these effects (Figure 2B).

6-aminocaproic acid (EACA) as a buffer in CE has a selectivity-enhancing effect, which is related to its ion-pair-forming property [33,34]. A significant improvement in resolution was observed compared to the analysis in acetate, although it did not provide baseline resolution (Figure 2C). To improve the selectivity for aldoses, an HPMC linear polymer was added to the 40 mM EACA buffer at a concentration of 0.02%. The cellulose derivative improved the resolution between glucose and mannose and extended the analysis time compared to the PEO-containing buffer (Figure 2D). The effect of HPMC in increasing the separation efficiency is similar to that of PEO. From among the BGEs investigated, 25 mM ammonium acetate-0.3% PEO 90,000 (pH 4.5) provided the best separation and the best resolution values (Supplementary Material Table S1); therefore, this BGE was applied for further analyses.

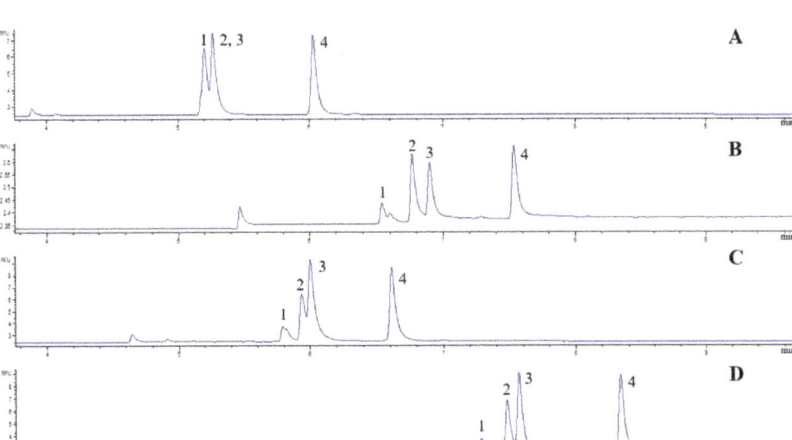

Figure 2. The CZE separations of a mixture of four carbohydrates using LEDIF detection. Conditions: fused silica capillary, l_{tot}: 65 cm × 50 µm id, l_{eff}: 45 cm, BGE: (**A**) 25 mM ammonium acetate (pH 4.5), (**B**) 25 mM ammonium acetate-0.3% PEO (pH 4.5), (**C**) 40 mM EACA (pH 4.5), (**D**) 40 mM EACA-0.02% HPMC (pH 4.5), separation voltage: −30 kV, injection: 50 mbar × 2 s, preconditioning: 1 M HCl for 5 min, BGE for 5 min washing. LEDIF detection was performed at 480 nm excitation/520 nm emission. Sample: 1: fructose, 2: glucose, 3: mannose, 4: maltose.

The use of PEO enabled the separation of positional isomers of oligosaccharides, as well. Two samples of maltotetraose (G4)- isomaltotetraose (isoG4) and maltotriose (G3)-isomaltotriose (isoG3) were analyzed (Figure 3A,B). In the case of G3 and G4, the monosaccharide units are connected by an α(1-4) bond, whereas in the isoG3 and isoG4, the position of the bond is different; the glucose units are linked by an α(1-6) bond. Since the electrophoretic mobility is also influenced by the shape of the particle, it was possible to separate the G3 and G4 from their positional isomers possessing the same mass and charge but different form. The α(1-4) linkage affords an elongated, thinner molecular shape; therefore the friction coefficients of G3 and G4 are higher, so their electrophoretic mobility is lower than that of their isomers. The spherical shape of the isomers led to lower friction coefficients and, as a result, higher migration speed (Figure 3A,B).

During method development, sample injection was also examined, since the honey samples were introduced into the capillary without any sample preparation procedure, except for the labeling reaction. Electrokinetic sample introduction of a standard carbohydrate solution with 5 kV × 5 s delivered the same amount of each carbohydrate into the capillary as hydrodynamic injection by 50 mbar × 2 s (Supplementary Material Figure S1). The ratio of individual carbohydrates did not vary when electrokinetic injection was applied because there is no difference in the charge of APTS-labeled sugars. It is possible to examine APTS-marked sugar components in samples of honey via electrokinetic injection.

The calibration curves and analytical performance data are given for two sugars (fructose and glucose) that are present in honey in the largest amounts (Table 1). The calculated LOD and limit of quantitation (LOQ) in the range of ng/mL enable a more sensitive determination than what is usually available with UV-VIS spectrophotometric detection in the case of CE (µg/mL). The relative standard deviation (RSD%) values of migration time were around 0.5, whereas RSD% values of area were between 2.5 and 4.4. The theoretical plate number data correspond to the separation efficiency expected from CE (Table 1).

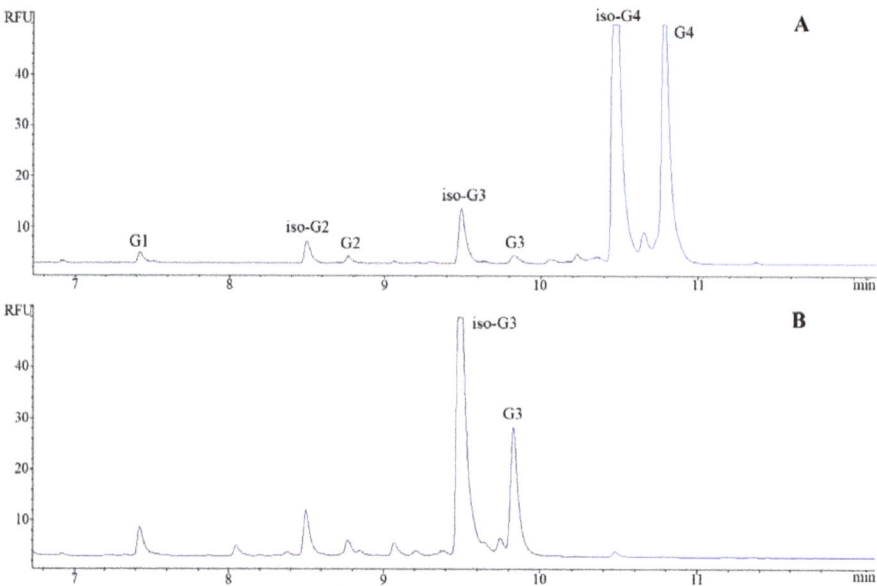

Figure 3. The analysis of isomaltotetraose (**A**) and isomaltotriose (**B**) by CZE-LEDIF. The analysis conditions were the same as stated at Figure 2B.

Table 1. Analytical performance data of fructose and glucose. The analysis conditions were the same as stated at Figure 2B.

		Glucose	Fructose
Regression equation		0.078x − 0.2736	0.044x − 11.39
Correlation coefficient		0.9988	0.9916
LOD (ng/mL)		8.89	8.76
LOQ (ng/mL)		29.6	29.2
Range (ng/mL)		30–1000	30–1000
RSD %, (N = 5)	time	0.496	0.509
	area	2.48	2.82
Theoretical plate number		110,140	215,657

3.2. Analysis of Sugars in Honey Samples

The sugar content of three different honey samples (phacelia honey, honeydew honey, and acacia honey) was analyzed by CE-LEDIF using ammonium acetate with a PEO additive as a BGE (Figure 4C–E). The sugar content of honeys was characterized using a carbohydrate ladder (Figure 4A) and a standard mixture containing five carbohydrates (arabinose, xylose, ribose, galactose and maltose) (Figure 4B). All three honeys contain two monosaccharides in higher concentrations, and other sugars appeared in smaller concentrations in the monosaccharide region (G1), according to the ladder (Figure 4C–E). This corresponds well with the literature data showing that honey consists of about 90% sugars, mainly glucose and fructose. Identification of glucose and fructose peaks in honeys was performed by standard addition. This is illustrated through the example of phacelia honey, wherein the addition of fructose resulted in an enhancement in the peak height of fructose, which is proportional to the added amount (Supplementary Material Figure S2). Apart from glucose and fructose, other sugars were not identified by spiking. The fructose and glucose content of honey in mass percentages and their relative proportion data were given using the prepared calibration curves (Table 2). The fructose-to-glucose ratio is one of the most important data for describing the quality of honeys [2,35]. Honey is crystallized when this ratio is less than 1.2. This data reveals which honey contains more glucose and indicates how likely it is to crystallize. Phacelia honey had the highest glucose content and the lowest fructose-to-glucose ratio. Honey is in a liquid state if the fructose-to-glucose ratio exceeds 1.2, so phacelia honey crystallized, whereas acacia and honeydew honeys retained their liquid characters due to the lower glucose content. Honeydew honey differed from phacelia and acacia honey in terms of monosaccharide composition. It had a lower carbohydrate content (Table 2). Honeydew honey is not of flower nectar origin, and, actually, this difference in origin is what causes the diversity in composition. Honeydew is excreted by aphids and other insects, and this excreted, sticky substance is collected by bees.

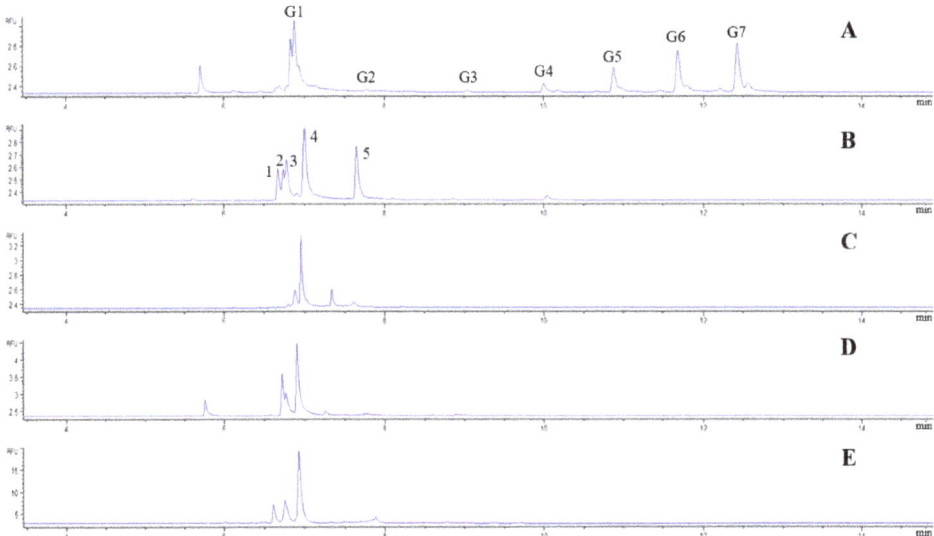

Figure 4. CZE-LEDIF determination of APTS-labeled sugars in model solutions and in honeys. (**A**) G7 malto-oligosaccharide ladder, (**B**) mixture solution of arabinose (1), xylose (2), ribose (3), galactose (4), maltose (5), (**C**) phacelia honey, (**D**) acacia honey, (**E**) honeydew honey. The analysis conditions were the same as stated at Figure 2B.

Table 2. The comparison of fructose and glucose content and their proportion in phacelia honey, acacia honey, honeydew honey, n = 3 (each of them). The analysis conditions were the same as stated at Figure 2B.

	Concentration (m/m%)		
	Phacelia honey	Acacia honey	Honeydew honey
Glucose	37.8 ± 1.5	26.7 ± 1.1	19 ± 0.76
Fructose	32.7 ± 1.3	43.2 ± 1.7	27.1 ± 1
Fructose/Glucose	0.86	1.61	1.42
n = 3			

In addition to monosaccharides, the presence of small amounts of disaccharides was confirmed in the honey samples using the carbohydrate ladder (Figure 4C–E). According to the literature, maltose, sucrose, maltulose, nigerose, kojibose, trehalose and turanose are the disaccharides that are most commonly found in honey [1,2]. None of the disaccharide peaks visible on the electropherogram come from sucrose; because sucrose is not a reducing disaccharide, it cannot be labeled by APTS. According to the carbohydrate ladder, the electropherograms between G1 and G2 showed the positional isomers of the disaccharides. The honeys analyzed contained very trace amounts of trisaccharides. Less than 0.1% of the total monosaccharide amount was made up by the peaks migrating at the G3 location.

The evaluation of the greenness of analytical methods was performed by the Analytical Greenness calculator (AGREE). The obtained score was 0.74, which is in the middle of the AGREE pictogram with values close to 1; the green color indicates that the applied CE-LEDIF analytical procedure has little environmental impact [36–38].

4. Conclusions

In this work, we studied the possibilities of determining the sugar composition of honey samples by capillary electrophoresis with light-emitting diode-induced fluorescence detection (CE-LEDIF). We applied 8-aminopyrene-1,3,6-trisulfonic acid (APTS) for the labeling of honey sugars, which made possible not only the sensitive detection but also the electrophoretic separation. LEDIF detection allowed for the determination of sugars with high sensitivity (~30 ng/mL limit of quantitation (LOQ) values). The effect of background electrolyte (BGE) additives on the separation efficiency was investigated, and the addition of polyethylene oxide (PEO) improved the resolution of labeled sugars having the same carbon number. We determined the glucose and fructose content of three different types of honey samples. The CE-LEDIF method has the potential to be a suitable platform for the routine analysis of honey samples in food and biology laboratories. Although the developed method is well suited for the determination of sugar components which are presented in honey, the method does not allow for the direct determination of sucrose, which plays a major role in the adulteration of honey. Further developments are needed for the determination of non-reducing disaccharides, for instance, sucrose (after enzymatic hydrolysis). We plan to analyze honey samples of different ages and different geographical origin.

Supplementary Materials: The following supporting information can be downloaded at: https://www.mdpi.com/article/10.3390/separations10030150/s1. Table S1: The comparison of resolutions using different BGEs. The analysis conditions were the same as stated in Figure 2. Figure S1: The CZE-LEDIF determinations of a mixture of four carbohydrates using hydrodynamic (A) and electrokinetic injection (B). The analysis conditions were the same as stated in Figure 2B. Injection: +5 kV × 5 s (S1B), samples: 1: fructose, 2: glucose, 3: mannose, 4: maltose, (*): APTS. Figure S2: Identification of fructose peak in honey using standard addition. The analysis conditions were the same as stated in Figure 2B. Sample: 1: fructose, 2: glucose, (*): APTS, (S2A) phacelia honey, (S2B) phacelia honey spiked with fructose, (S2A', S2B') narrow scale of electropherograms of (A, B).

Author Contributions: Conceptualization, M.A., G.G. and Z.S.; methodology, M.A.; investigation, M.A.; writing—original draft preparation, M.A.; writing—review and editing, C.N.; supervision, G.G., Z.S. and C.N. All authors have read and agreed to the published version of the manuscript.

Funding: The research presented in the article was carried out within the framework of the National Research, Development and Innovation Office (K127931 and K142134), Hungary.

Data Availability Statement: Data available upon request from the authors.

Acknowledgments: The authors acknowledge the financial support provided to this project by the National Research, Development and Innovation Office (K127931 and K142134), Hungary.

Conflicts of Interest: The authors declare no conflict of interest.

Abbreviations

capillary electrophoresis (CE); light-emitting diode-induced fluorescence detection (LEDIF); polyethylene oxide (PEO); 8-aminopyrene-1,3,6-trisulfonic acid (APTS); ultraviolet-visible (UV-VIS); background electrolyte (BGE); acid dissociation constant (pKa); capillary zone electrophoresis (CZE); mass spectrometry (MS); limit of detection (LOD); laser-induced fluorescence detection (LIF); 8-aminonaphthalene-1,3,6-trisulfonic acid (ANTS); aminonaphthalene-2-sulfonic acid (ANA); 4-amino-5-hydroxynaphthalene-2,7-disulfonic acid (AHNS); cetyltrimethylammonium bromide (CTAB); sodium hydroxide (NaOH); 1-phenyl-3-methyl-5-pyrazolone (PMP); graphene–cobalt microsphere (CoMS); hydrochloric acid (HCl); 6-aminocaproic acid (EACA); hydroxypropyl methylcellulose (HPMC); sodium cyanoborohydride (NaBH$_3$CN); tetrahydrofuran (THF); signal-to-noise ratio (S/N); numbers of glucose residues of the malto-oligosaccharide ladder standard (GX); maltotetraose (G4); isomaltotetraose (isoG4); maltotriose (G3); isomaltotriose (isoG3); limit of quantitation (LOQ); relative standard deviation (RSD%).

References

1. Samarghandian, S.; Farkhondeh, T.; Samini, F. Honey and health: A review of recent clinical research. *Phcog. Res.* **2017**, *9*, 121–127. [PubMed]
2. Machado de-Melo, A.A.; Bicudo de Almeida-Muradian, L.; Sancho, M.T.; Pascual-Mate, A. Composition and properties of Apis mellifera honey: A review. *J. Apic. Res.* **2018**, *57*, 5–37. [CrossRef]
3. Asquieri, E.R.; Gomes de Moura, A.S.; Dianiny de Cassia, S.M.; Batista, R.D. Comparison of titulometric and spectrophotometric approaches towards the determination of total soluble and insoluble carbohydrates in foodstuff. *Carpathian J. Food Sci. Technol.* **2019**, *11*, 69–79.
4. BeMiller, J.N. *Chapter 10-Carbohydrate Analysis, Food Analysis*; Springer: Berlin/Heidelberg, Germany, 2010; pp. 147–177.
5. Slimestad, R.; Vagen, I.M. Thermal stability of glucose and other sugar aldoses in normal phase high performance liquid chromatography. *J. Chromatogr. A* **2006**, *1118*, 281–284. [CrossRef]
6. Cordella, C.; Militao, J.S.L.T.; Clement, M.C.; Drajnudela, P.; Cabrol-Bass, D. Detection and quantification of honey adulteration via direct incorporation of sugar syrups or bee-feeding: Preliminary study using high-performance anion exchange chromatography with pulsed amperometric detection (HPAEC-PAD) and chemometrics. *Anal. Chim. Acta* **2005**, *531*, 239–248. [CrossRef]
7. Hetricka, E.M.; Kramer, T.T.; Risleya, D.S. Evaluation of a hydrophilic interaction liquid chromatography design space for sugars and sugar alcohols. *J. Chromatogr. A* **2017**, *1489*, 65–74. [CrossRef]
8. Weiβ, K.; Alt, M. Determination of single sugars, including inulin, in plants and feed materials by high-performance liquid chromatography and refraction index detection. *Fermentation* **2017**, *3*, 36.
9. Al-Sanea, M.M.; Gamal, M. Critical analytical review: Rare and recent applications of refractive index detector in HPLC chromatographic drug analysis. *Microchem. J.* **2022**, *178*, 107339. [CrossRef]
10. Jalaludin, I.; Kim, J. Comparison of ultraviolet and refractive index detections in the HPLC analysis of sugars. *Food Chem.* **2021**, *365*, 130514. [CrossRef]
11. Suzuki, S.; Honda, S. A tabulated review of capillary electrophoresis of carbohydrates. *Electrophoresis* **1998**, *19*, 2539–2560. [CrossRef]
12. Rassi, Z.E. Recent developments in capillary electrophoresis and capillary electrochromatography of carbohydrate species. *Electrophoresis* **1999**, *20*, 3134–3144. [CrossRef]
13. Lu, G.; Crihfield, C.L.; Gattu, S.; Veltri, L.M.; Holland, L.A. Capillary electrophoresis separations of glycans. *Chem. Rev.* **2018**, *118*, 7867–7885. [CrossRef] [PubMed]
14. Arentoft, A.M.; Michaelsen, S.; Sorensen, H. Determination of oligosaccharides by capillary zone electrophoresis. *J. Chromatogr. A* **1993**, *652*, 517–524. [CrossRef]
15. Honda, S. Separation of neutral carbohydrates by capillary electrophoresis. *J. Chromatogr. A* **1996**, *720*, 337–351. [CrossRef]
16. Mantovani, V.; Galeotti, F.; Maccari, F.; Volpi, N. Recent advances in capillary electrophoresis separation of monosaccharides, oligosaccharides, and polysaccharides. *Electrophoresis* **2018**, *39*, 179–189. [CrossRef]
17. Soga, T.; Serwe, M. Determination of carbohydrates in food samples by capillary electrophoresis with indirect UV detection. *Food Chem.* **2000**, *69*, 339–344. [CrossRef]
18. Kakehia, K.; Funakuboa, T.; Suzukia, S.; Odaa, Y.; Kitada, Y. 3-Aminobenzamide and 3-aminobenzoic acid, tags for capillary electrophoresis of complex carbohydrates with laser-induced fluorescent detection. *J. Chromatogr. A* **1999**, *863*, 205–218. [CrossRef]
19. Carvalho, A.Z.; da Silva, J.A.F.; do Lago, C.L. Determination of mono- and disaccharides by capillary electrophoresis with contactless conductivity detection. *Electrophoresis* **2003**, *24*, 2138–2143. [CrossRef]
20. Campa, C.; Coslovi, A.; Flamigni, A.; Rossi, M. Overview on advances in capillary electrophoresis-mass spectrometry of carbohydrates: A tabulated review. *Electrophoresis* **2006**, *27*, 2027–2050. [CrossRef]
21. Starr, C.M.; Masada, R.I.; Hague, C.; Skop, E.; Klock, J.C. Fluorophore-assisted carbohydrate electrophoresis in the separation, analysis, and sequencing of carbohydrates. *J. Chromatogr A* **1996**, *720*, 295–321. [CrossRef]
22. Paulus, A.; Klockow, A. Detection of carbohydrates in capillary electrophoresis. *J. Chromatogr. A* **1996**, *720*, 353–376. [CrossRef] [PubMed]
23. Guttman, A. Analysis of monosaccharide composition by capillary electrophoresis. *J. Chromatogr. A* **1997**, *763*, 271–277. [CrossRef] [PubMed]
24. Olajos, M.; Hajos, P.; Guenther, K.; Bonn, G.K.; Guttman, A. Sample preparation for the analysis of complex carbohydrates by multicapillary gel electrophoresis with light emitting diode induced fluorescence detection. *Anal. Chem.* **2008**, *80*, 4241–4246. [CrossRef] [PubMed]
25. Shaheen, R.; Senn, J.P. Quantification of polysaccharides in water using capillary electrophoresis. *Int. J. Environ. Anal. Chem.* **2005**, *85*, 177–198. [CrossRef]
26. Rizelio, V.M.; Tenfen, L.; Silveira, R.; Valdemiro, L.; Gonzaga, L.V.; Costa, A.C.O.; Fett, R. Development of a fast capillary electrophoresis method for determination of carbohydrates in honey samples. *Talanta* **2012**, *93*, 62–66. [CrossRef]

27. Dominguez, M.A.; Jacksen, J.; Emmer, A.; Centurion, M.E. Capillary electrophoresis method for the simultaneous determination of carbohydrates and proline in honey samples. *Microchem. J.* **2016**, *129*, 1–4. [CrossRef]
28. Lua, Y.; Hub, Y.; Wangc, T.; Yanga, X.; Zhaoc, Y. Rapid determination and quantitation of compositional carbohydrates to identify honey by capillary zone electrophoresis. *J. Food* **2017**, *15*, 531–537. [CrossRef]
29. Liang, P.; Sun, M.; He, P.; Zhang, L.; Chen, G. Determination of carbohydrates in honey and milk by capillary electrophoresis in combination with graphene–cobalt microsphere hybrid paste electrodes. *Food Chem.* **2016**, *190*, 64–70. [CrossRef]
30. Chen, F.T.; Dobashi, T.S.; Evangelista, R.A. Quantitative analysis of sugar constituents of glycoproteins by capillary electrophoresis. *Glycobiology* **1998**, *8*, 1045–1052. [CrossRef]
31. Volpi, N. *Capillary Electrophoresis of Carbohydrates, from Monosaccharides to Complex Polysaccharides*; Springer: Berlin/Heidelberg, Germany, 2011.
32. Horvath, J.; Dolnik, V. Polymer wall coatings for capillary electrophoresis. *Electrophoresis* **2001**, *22*, 644–655. [CrossRef]
33. Hamm, M.; Wang, Y.; Rustandi, R.R. Characterization of N-Linked glycosylation in a monoclonal antibody produced in NS0 cells using capillary electrophoresis with laser-induced fluorescence detection. *Pharmaceuticals* **2013**, *6*, 393–406. [CrossRef] [PubMed]
34. Bunz, S.C.; Rapp, E.; Neususs, C. Capillary electrophoresis/mass spectrometry of APTS-labeled glycans for the identification of unknown glycan species in capillary electrophoresis/laser-induced fluorescence systems. *Anal. Chem.* **2013**, *85*, 10218–10224. [CrossRef] [PubMed]
35. Manikis, I.; Thrasivoulou, A. The relation of physicochemical characteristics of honey and the crystallization sensitive parameters. *Apiacta* **2001**, *36*, 106–112.
36. Gamal, M.; Naguib, I.A.; Panda, D.S.; Abdallah, F. Comparative study of four greenness assessment tools for selection of greenest analytical method for assay of hyoscine N-butyl bromide. *Anal. Methods* **2021**, *13*, 369–380. [CrossRef]
37. Pena-Pereira, F.; Wojnowski, W.; Tobiszewski, M. AGREE—Analytical GREEnness metric approach and software. *Anal. Chem.* **2020**, *92*, 10076–10082. [CrossRef]
38. Plotka-Wasylka, J.; Wojnowsk, W. Complementary green analytical procedure index (ComplexGAPI) and software. *Green Chem.* **2021**, *23*, 8657–8665. [CrossRef]

Disclaimer/Publisher's Note: The statements, opinions and data contained in all publications are solely those of the individual author(s) and contributor(s) and not of MDPI and/or the editor(s). MDPI and/or the editor(s) disclaim responsibility for any injury to people or property resulting from any ideas, methods, instructions or products referred to in the content.

 separations

Article

UPLC-MS/MS Method for Simultaneous Estimation of Neratinib and Naringenin in Rat Plasma: Greenness Assessment and Application to Therapeutic Drug Monitoring

Ali Altharawi [1,*], Safar M. Alqahtani [1], Sagar Suman Panda [2], Majed Alrobaian [3], Alhumaidi B. Alabbas [1], Waleed Hassan Almalki [4], Manal A. Alossaimi [1], Md. Abul Barkat [5], Rehan Abdur Rub [6], Shehla Nasar Mir Najib Ullah [7], Mahfoozur Rahman [8] and Sarwar Beg [6]

1. Department of Pharmaceutical Chemistry, College of Pharmacy, Prince Sattam Bin Abdulaziz University, Al-Kharj 11942, Saudi Arabia
2. Department of Pharmaceutical Analysis and Quality Assurance, Roland Institute of Pharmaceutical Sciences, Berhampur 760010, India
3. Department of Pharmaceutics and Industrial Pharmacy, College of Pharmacy, Taif University, Taif 21944, Saudi Arabia
4. Department of Pharmacology and Toxicology, College of Pharmacy, Umm Al-Qura University, Makkah 24381, Saudi Arabia
5. Department of Pharmaceutics, College of Pharmacy, University of Hafr Al Batin, Hafar Al Batin 39524, Saudi Arabia
6. Department of Pharmaceutics, School of Pharmaceutical education and Research, Jamia-Hamdard, Hamdard University, New Delhi 110062, India
7. Department of Pharmacognosy, Faculty of Pharmacy, King Khalid University, Abha 62529, Saudi Arabia
8. Department of Pharmaceutical Sciences, Shalom Institute of Health & Allied Sciences, Sam Higginbottom University of Agriculture, Technology & Sciences, Allahabad 211007, India
* Correspondence: a.altharawi@psau.edu.sa

Abstract: Tyrosine kinase inhibitors have often been reported to treat early-stage hormone-receptor-positive breast cancers. In particular, neratinib has shown positive responses in stage I and II cases in women with HER2-positive breast cancers with trastuzumab. In order to augment the biopharmaceutical attributes of the drug, the work designed endeavors to explore the therapeutic benefits of neratinib in combination with naringenin, a phytoconstituent with reported uses in breast cancer. A UPLC-MS/MS method was developed for the simultaneous estimation of neratinib and naringenin in rat plasma, while imatinib was selected as the internal standard (IS). Acetonitrile was used as the liquid extractant. The reversed-phase separation was achieved on a C18 column (100 mm × 2.1 mm, 1.7 µm) with the isocratic flow of mobile phase-containing acetonitrile (0.1% formic acid) and 0.002 M ammonium acetate (50:50, % v/v) at flow rate 0.5 mL·min^{-1}. The mass spectra were recorded by multiple reaction monitoring of the precursor-to-product ion transitions for neratinib (m/z 557.138→111.927), naringenin (m/z 273.115→152.954), and the IS (m/z 494.24→394.11). The method was validated for selectivity, trueness, precision, matrix effect, recovery, and stability over a concentration range of 10–1280 ng·mL^{-1} for both targets and was acceptable. The method was also assessed for greenness profile by an integrative qualitative and quantitative approach; the results corroborated the eco-friendly nature of the method. Therefore, the developed method has implications for its applicability in clinical sample analysis from pharmacokinetic studies in human studies to support the therapeutic drug monitoring (TDM) of combination drugs.

Keywords: breast cancer; liquid chromatography; bioanalytical methods; neratinib; naringenin

1. Introduction

Breast cancer is considered one of the leading causes of mortality in the female population and the second most commonly reported cancer globally [1]. Over recent years, there has been continuous drug approval for the treatment and management of breast cancer. As per the WHO statistics of 2020, around 2.5 million deaths are estimated between 2020 and 2040. The early diagnosis of breast cancer, with confirmation of its genetic basis and heterogeneity, helps in selecting the right chemotherapeutic drugs for breast cancer treatment. In the past few years, the United States Food Drug Administration (USFDA) has approved several tyrosine kinase inhibitors for breast cancer treatment, which are very effective in HER2-overexpressed/amplified breast cancers [2,3]. Tyrosine kinase inhibitors specifically block the abnormal signal transduction, which is necessary for the proliferation of the cancer cells, and also show anti-epithelial growth factor receptor (EGFR) activity.

Neratinib (NER), a tyrosine kinase inhibitor, was approved by the USFDA in July 2017 as an extended adjuvant therapy for adult patients prescribed with trastuzumab for early-stage breast cancers [4,5]. Moreover, it is available in the form of tablets for oral administration at a dose of 40 mg. In contrast, the oral pharmacokinetics of neratinib exhibit a nonlinear absorption profile along with a food effect [6]. To understand the pharmacokinetic and biodistribution profile of the drug, a fast, sensitive, and efficient analytical method is always very useful for preclinical and clinical sample analysis. Very few bioanalytical methods of neratinib alone have been reported in the literature for estimation in rat and human plasma using high-performance liquid chromatography (HPLC) with ultraviolet (UV) and diode/photodiode array detectors (DAD/PDA) [7] and ultra-performance liquid chromatography (UPLC) with mass spectrometer (MS/MS) detectors [8–10]. In contrast to these existing methods, the present work aimed to establish a rapid and sensitive bioanalytical method for the quantification of multiple compounds in human plasma samples. This new method also endeavors to support the therapeutic drug monitoring (TDM) program. LC-MS/MS methods have been quite frequently used in TDM owing to their high sensitivity and specificity compared to other analytical techniques, as they quantify drugs irrespective of their natural chromophores or fluorophores [11,12]. Upon the optimization of sample preparation and mass spectrometry conditions, column and internal standard selection, LC-MS/MS methods greatly help in avoiding interference due to the matrix effect and the presence of other analytes and metabolites [13,14].

For better pharmacotherapeutic action, chemotherapeutic drug combinations with phytopharmaceuticals, especially antioxidants, have been extensively explored in the past few decades [15,16]. In this regard, we found naringenin (NRN) to be one of the potential phytopharmaceutical agents for various cancers, including breast cancer [17,18]. NRN potentially exhibits a dose-dependent increase in caspase-3 and caspase-9 activity-mediated apoptosis in breast cancer cells [17]. Additionally, it reduces cell proliferation and inhibits the migration of breast cancer cells via inflammatory and apoptosis cell signaling pathways. For the quantification of NRN in preclinical and clinical samples, a suitable bioanalytical method is required to understand the pharmacokinetic and pharmacodynamic behaviors of the drug. A score of research studies have documented the analytical estimation of NRN in biological matrices employing HPLC [19–21], LC-MS/MS [22], and related techniques. However, the sensitivity or lower limit of quantification (LLOQ) reported in the HPLC methods was in the range of 0.1–15 $\mu g \cdot mL^{-1}$ [19–21], while the LC-MS/MS method by Ma et al. (2006) reported double peaks for NRN in rat plasma despite a good LLOQ of 5 $ng \cdot mL^{-1}$ [22]. On the contrary, the current method showed a single peak for NRN with high resolution, although LLOQ was found to be 10 $ng \cdot mL^{-1}$, which is still good for the routine analysis of the drugs in preclinical and clinical samples.

Green analytical chemistry (GAC) is a recent approach growing in popularity amongst analysts. However, any analytical procedure's green or eco-friendly nature dramatically affects environmental sustainability and the overall economy of the method development. With many screening and identification processes, chemists have recommended solvents and reagents that promise to fulfill the above intent. With scientists presenting newer approaches now and then, it has been a critical task for the analyst to choose between the various greenness assessment tools available. The National Environmental Methods Index (NEMI), Green Analytical Procedure Index (GAPI), and eco-scale (etcetera) are some of the most widely utilized tools for assessing a method's greenness. Several scientists have reported the benefits and justifications of using such approaches for examining the eco-friendly nature of different analytical procedures [23,24]. A detailed discussion of their goals and rationale for assessment can be found elsewhere. In the current study, two assessment approaches, namely NEMI and AES, were adopted to qualitatively and semi-quantitatively determine the method greenness score.

In analytical science, method development for simultaneous estimation of two or more compounds provides cost and time economy in the analysis and colossal resource savings. However, method development for the simultaneous estimation of compounds is quite tedious, as many factors tend to influence the retention capacity of the drugs due to variations in their physicochemical characteristics. Therefore, in the present work, the researchers focused on developing a UPLC-MS/MS method to estimate NER and NRN in rat plasma simultaneously. Furthermore, the developed analytical method was validated according to current regulatory guidelines and employed to assess the stability of both analytes [25]. Additionally, the bioanalytical method's greenness was assessed considering some of the latest approaches discussed by various analysts [23,24].

2. Materials and Methods

2.1. Chemicals and Reagents

NER was purchased from Weihua Pharma (Hangzhou, China), and NRN (Purity > 99%) was purchased from TCI Chemicals (India) Pvt. Ltd. (Chennai, India), while the internal standard (IS), imatinib (Purity > 98%), was generously provided by Dr. Reddy's Laboratories Ltd. (Hyderabad, India). The LC-MS grade acetonitrile (ACN) and distilled water were purchased from J.T. Baker Chemicals (Mumbai, India), and ammonium acetate, ethyl acetate, formic acid, and ammonium formate were obtained from Fluka Analytical (Seelze, Germany). The other chemicals and reagents obtained were of analytical reagent grade.

2.2. UPLC-MS/MS Instrument and Conditions

The instrumental analysis was performed on the ACQUITY® UPLC-MS/MS system (Waters Corp., Milford, MA, USA) fitted with a binary pump system, autosampler unit, and column compartment. A Zspray™ Xevo TQD (Waters Corp., Milford, MA, USA) mass spectrometer working with positive mode electrospray ionization (ESI) detected and quantified the analytes of NER and NRN. The detailed instrumental setting preferences for different test parameters are displayed in the Supplementary Information, Table S1.

2.3. Animal Ethical Approval

The animal study protocol was approved by the Institutional Animal Ethic Committee (IAEC) of Roland Institute of Pharmaceutical Sciences (RIPS) (Berhampur, Odisha 760010, India), with the protocol number 146/Chairman IAEC, RIPS, Berhampur (Approval Date: 13 November 2021). The animal care and maintenance were carried out as per the guidelines of the Committee for the Purpose of Control and Supervision of Experiments on Animals (CPCSEA). Healthy male Wistar rats (weighing between 180 and 220 g) were housed in polypropylene cages with free access to a standard diet and water *ad libitum*. Moreover, the animals were exposed to regular day and night cycles to maintain their circadian rhythms.

2.4. Preparation of Standards and Sample

Separate stock solutions (1 mg·mL^{-1}) of NER, NRN, and IS were prepared using ACN as a diluent. Precisely measured, 0.05 mL and 0.1 mL of analytes and IS were pipetted from the stocks and spiked into rat plasma samples (0.5 mL). After liquid-liquid extraction (LLE), the final dilution of eight concentrations of 10, 20, 40, 80, 160, 320, 640, and 1280 ng·mL^{-1} of NER and NRN and 10 ng·mL^{-1} of IS were obtained. Then, three specified concentrations viz. 100, 500, and 1000 ng·mL^{-1} were defined as QC solutions for the analytes, while

10 ng·mL^{-1} was taken as the IS for the validation results. The solutions were stored at −20 °C until the final analysis.

2.5. Bioanalytical Extraction from the Rat Plasma

Blood was collected from the retro-orbital plexus of anesthetized Wistar rats. The heparinized blood produced plasma when centrifuged at 10,000 rpm for 10 min. The recovered plasma was kept at −20 °C after the plasma harvesting operation. Aliquots of 0.05 mL of the analytes and 0.1 mL of 200 ng·mL^{-1} IS were accurately measured and spiked in Eppendorf tubes, which were well mixed for 5 min. The extraction medium, ACN (1 mL), was added to all samples, followed by 2 min of vortex mixing and 5 min centrifugation at 3000 rpm (4 ± 2 °C). Precisely measured 0.5 mL of the organic fraction was pipetted and dried. The final dilution was prepared to 1 mL using mobile the phase, and the syringe was filtered before injection into the instrument. The prepared samples were always stored at −20 °C until analysis.

2.6. Validation Protocol

Multiple bioanalytical method validation guidelines, such as the USFDA, China Food and Drug Administration (CFDA), etc., were referred to and employed to establish the validity of the current UPLC-MS/MS method.

2.6.1. Selectivity

Selectivity plays a critical role in establishing the uninterefered simultaneous quantification of analytes in a biological matrix, such as plasma. The absence of interference at analyte retention was used to establish method selectivity by comparing the chromatograms of the blank plasma from six animals with spiked samples.

2.6.2. Linearity and Sensitivity

Over the course of three days, eight-point analyte linearity curves were created utilizing the spiked plasma samples with concentrations ranging from 10 to 1280 ng·mL^{-1}. The linearity plots were created using the drug-to-IS peak area ratio (y-axis) versus the spiked NER and NRN concentrations (x-axis). Further, the lower limit of quantification (LLOQ) was determined using six replicate samples that qualify for a minimum level of trueness and precision. For establishing trueness and precision at LLOQ, a maximum allowable deviation of 20% was permitted.

2.6.3. Trueness and Precision

We conducted hexaplicate analysis at LLOQ (10 ng·mL^{-1}) and low, mid, and high QC samples with 100, 500, and 1000 ng·mL^{-1} of NER and NRN, respectively, over three days to examine the trueness and precision of the method. The authors used percent relative standard deviation (% RSD) to define the method's trueness and precision, in addition to the amount of analyte recovered. The maximum variance allowed was 15% of the specified nominal level.

2.6.4. Carryover and Dilution Integrity

A blank sample was injected after the immediate analysis of the highest calibration standard to test carry-over. The absence of a significant carry-over can be confirmed by obtaining a peak area not more (<20%) than that of the LLOQ.

Subsequently, the dilution integrity was tested by spiking the blank rat plasma with analytes and diluting it ten times in blank plasma within the studied range, with a maximum allowed deviation of ±15% for trueness and precision.

2.6.5. Matrix Effect and Extraction

The extracted peak regions of NER, NRN, and IS were directly compared to the spiked plasmatic blanks, revealing the extraction recovery at three QC levels. Additionally, a comparison of the peak areas of the analytes and IS in the plasma of rats to that of the equi-concentration standard solutions leads to inferring the extent of the influence of the matrix.

2.6.6. Stability Study

NRB and NRN were studied for their stability in bench top and short-term (room temperature) for 6 h and 24 h, respectively. At a temperature of $-20\,°C$, long-term (14 days) and freeze–thaw (3 cycles) stability was assessed. The post-preparative stability study of the analytes was performed at 8 °C. All of the above stability studies were performed using analytes at their LQC and HQC concentration levels (n = 6).

2.7. Integrative Application of Multiple Green Metrics Tools

The integrated approach of combining the National Environmental Methods Index (NEMI) procedure and the Analytical Eco-scale (AES) method was followed to ensure the greenness of the developed bioanalytical method [24,25]. NEMI deals with the identification of chemicals and reagents that are harmful to the environment and assigns green shades to eco-friendly ones in a typical pictogram highlighting each of them. Such reagents and chemicals are broadly classified as PBT (persistent, bioaccumulative, and toxic), H (hazardous), C (corrosive), and W (waste generation capability). Contrary to NEMI, a purely semi-quantitative approach of AES allocates penalty points (PPs) to an analytical method according to a corresponding amount of reagent depleted and subsequent possible hazards per single analysis. Additionally, it considers overall power demand, waste produced, and any operational risks for assigning PPs. Final greenness scores of 75 and above identify a method as being eco-friendly [26–28].

3. Results

3.1. Preparation of the Plasma Samples

A single-step extraction procedure was used to extract NER, NRN, and the IS from the rat plasma. The procedure employed acetonitrile as the most suitable extraction medium with higher recoveries. The addition of acetonitrile helped in protein precipitation and the partition of the analytes and IS during the liquid–liquid extraction process. The extraction recovery was observed to be quite satisfactory for the investigated analytes and the IS. There were no other peaks observed for the plasma components except for the analytes and IS, which ratified a lack of interference with the analytes.

3.2. Mobile Phase Selection and Optimization: A Greenness Perspective

The mobile phase of selection for a UPLC-MS/MS relies solely on the volatility properties and compatibility with the MS/MS system. Only a few restricted options befit these criteria and are used accordingly for vivid applications. However, if a green UPLC-MS/MS method is intended for development, one must adhere to the principles of green analytical chemistry (GAC) and prioritize the use of ethanol as an organic proportion. However, the greenest solvent, ethanol, is less preferred in reversed-phase chromatography because of higher viscosity and UV cutoff than methanol or acetonitrile. With lower viscosity and UV cutoff values than methanol, acetonitrile is the most preferred organic phase for UPLC-MS/MS. Additionally, it has better volatility properties than methanol, which support its use as a mobile phase component. Adding 0.1% formic acid into the mobile phase supports investigations under the positive ionization mode. Ammonium acetate in LC-MS grade water with a pH adjusted to 3.5 using acetic acid served as the aqueous phase. Finally, the method's greenness was ensured by executing the NEMI and AES work strategies, which are described below in detail.

3.3. Investigation of Method Greenness

For a preliminary qualitative method greenness evaluation, the NEMI pictogram procedure was executed. In this procedure, the persistent, bioaccumulative, toxic, hazardous, and corrosive chemicals and reagents and their probable waste production capabilities were assumed to be alarming if they crossed 50 g. In the present evaluation, three quadrants indicated greenness, except for the quadrant denoted for hazard (H). Figure 1 portrays the NEMI multi-quadrant plot with notions for associated hazards. This helped us to be cautious for minimal possible hazards and to proceed with the penalty scoring system of AES. An overall score of 90 (Table 1) categorized the present method as green. This integrated approach ensured method greenness. From the above investigations, we inferred the exceptionally green nature of the method and its suitability for routine use, as it supports the current thinking to establish an AES framework.

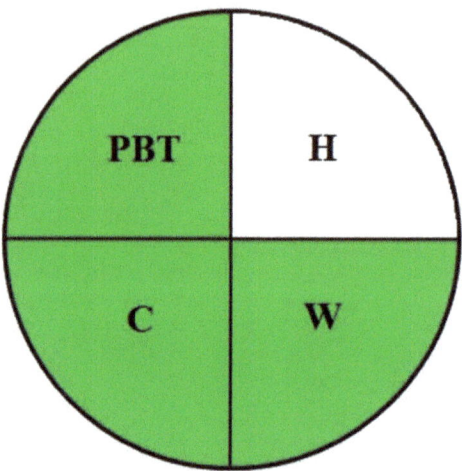

Figure 1. NEMI-oriented method greenness assessment pictogram.

Table 1. Overall penalty points and greenness score of the current bioanalytical method.

Reagents	Amount	Amount PP [a]	Hazard	Hazard PP	Total PP [b]
(1) ACN	<10 mL	1	Yes	4	4
(2) Acetate buffer	<10 mL	1	None	0	0
Instrument		Energy			
(1) UPLC-MS/MS		1.5 kWh per sample			2
(2) Ultrasonicator bath		<0.1 kWh per sample			0
(3) Filtration pump		<0.1 kWh per sample			0
(4) Centrifuge Unit		≤1.5 kWh per sample			1
(5) Occupational hazards		Process Hermetization			0
(6) Waste		<10 mL			3
Overall PPs:					10
AES total score:					100 − 10 = 90
Remark:					Excellent greenness

[a] Penalty points. [b] Total PP = Amount PP × Hazard PP.

3.4. Optimized LC and MS Conditions

ACQUITY UPLC® ethylene bridged hybrid (BEH) columns (Waters Corp., Milford, MA, USA) made of 1.7 µm particles of C8 and C18 were tested. Upon testing with the isocratic mode of separation, desired chromatographic results were obtained within 4 min. The flow of the mobile phase was 0.5 mL·min^{-1} (Supporting Information, Table S1). In-built Intellistart functionality automated the MS scanning and ESI optimization for NER,

NRN, and the IS. This function selected the ion transitions (Figure 2A–C) for NER (m/z of 557.138→111.927), NRN (m/z of 273.115→152.954), and the IS (m/z 494.5→394) out of the three available peaks. The Intellistart supported automated optimized MS/MS conditions for the valid quantification of the analytes, which are listed in the Supporting Information, Table S2. The typical MS/MS fragmentation products for NER, NRN, and the IS have also been provided in Figure 2.

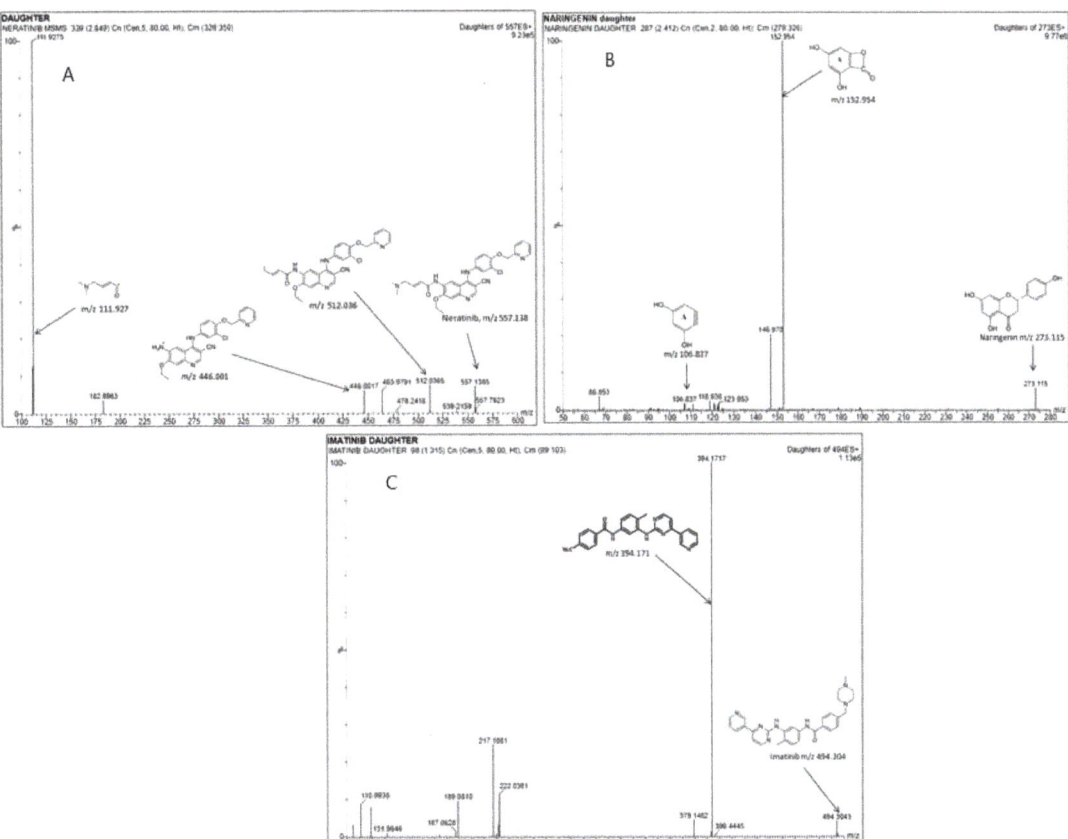

Figure 2. Representative parent to daughter ion MRM transitions and the fragmentation products formed for (**A**) NER, (**B**) IS, and (**C**) NRN at LLOQ.

3.5. Validation of Results

The results for the method validation parameters and their acceptance criteria are described in the below sections.

3.5.1. Method Selectivity

The analyzed MRM chromatograms recorded for the blank plasma of six separate animals (Supporting Information, Figure S1A–C) showed an absence of any interference at the identified retention times for NER and NRN (Supporting Information, Figure S2A–C). This ensured adequate method selectivity to support the quantification of the analytes.

3.5.2. Method Linearity and Quantitation Limits

A concentration range over 10–1280 ng·mL^{-1} of both NER (y = 0.082x + 0.071) and NRN (y = 0.060x + 0.058) was found to be linear (R^2 = 0.998). Figure 3 depicts the linear calibration plot for NER and NRN between concentrations (ng·mL^{-1}) versus the analyte/IS peak area ratio. At the lowest concentration of 10 ng·mL^{-1}, both the analytes (Table 2) were adequately quantified (Figure 4A–C) and established as the method's LLOQ values.

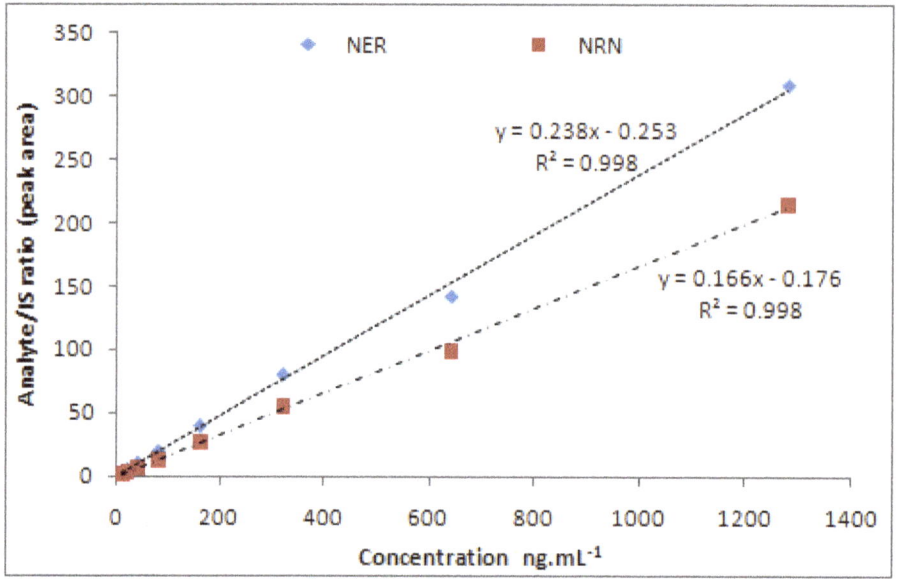

Figure 3. Calibration curve of NER and NRN.

Table 2. Method trueness and precision data.

Analyte	Nominal Concentration (ng·mL^{-1})	Intraday (n = 6)		Interday (n = 18)	
		Trueness (RE, %)	Precision (RSD, %)	Trueness (RE, %)	Precision (RSD, %)
NER	10 (LLOQ)	−10.5	2.4	−10.1	2.2
	100 (LQC)	−9.2	2.1	−10.2	2.5
	500 (MQC)	−8.7	1.4	−9.1	1.2
	1000 (HQC)	−10.23	1.5	−10.5	1.3
NRN	10 (LLOQ)	−10.5	1.8	−9.3	1.8
	100 (LQC)	−8.76	2.8	−10.6	1.6
	500 (MQC)	−10.36	1.4	−9.91	1.26
	1000 (HQC)	−10.02	1.3	−10.01	1.1

RE: Relative error; RSD: Relative standard deviation.

3.5.3. Trueness and Precision

The results obtained for trueness (% recovery) and precision (% RSD) at LLOQ, LQC, MQC, and HQC levels are listed in Table 2. In addition, the intraday and interday recoveries for NER were found to be between −8.7 and −10.5% (% RE), while for NRN, the values were found to be between −8.76 and −10.6% (% RE), respectively. Moreover, the method's preciseness was found to be between 1.1 and 2.8% (% RSD) for both the analytes, indicating its acceptability for bioanalytical applications.

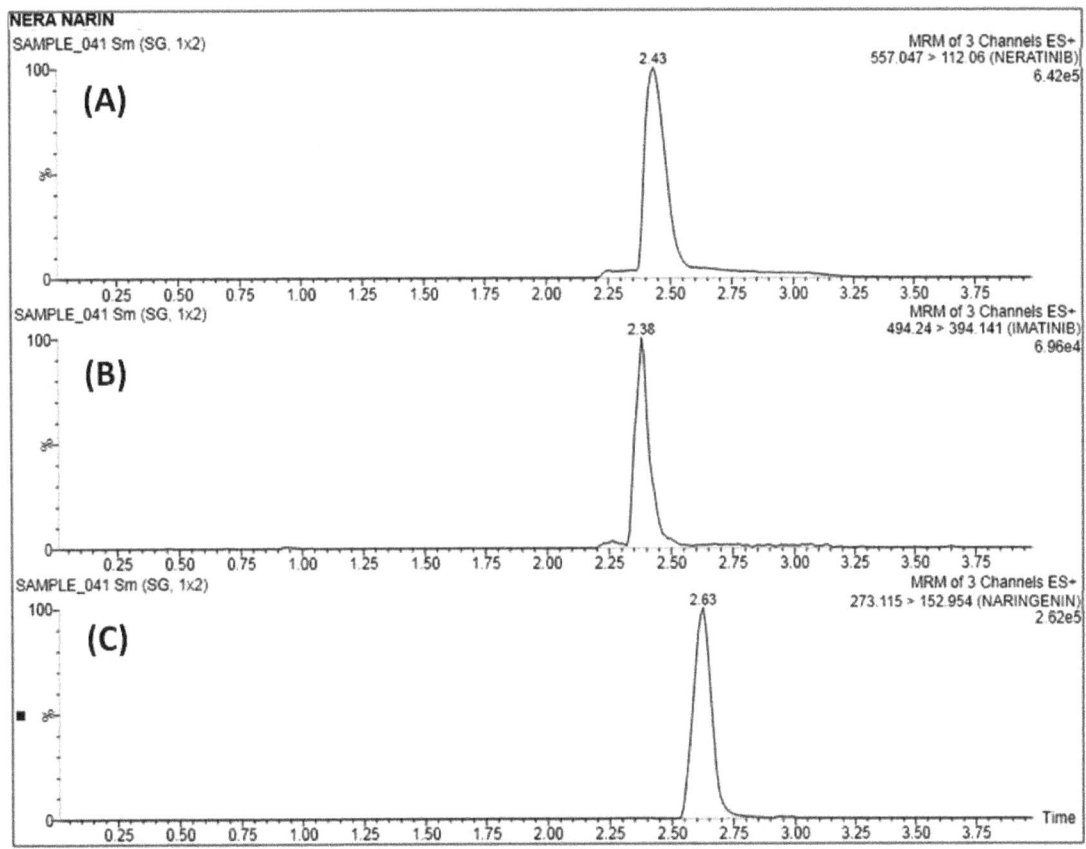

Figure 4. Representative MRM chromatogram of blank plasma samples of (**A**) NER, (**B**) IS, and (**C**) NRN at LLOQ.

3.5.4. Carryover and Dilution Integrity

Carryover was not observed by analyzing the blank plasma samples instantly after analyzing the highest calibration concentration of the analytes. In the dilution integrity studies for NER and NRN, the actual samples' concentrations were above the upper limit of the calibration range. The mean diluted concentrations were acceptable within ±6.8 and 7.5% of the nominal concentration for NER and NRN, respectively.

3.5.5. Extraction Recovery and Matrix Effect

Analyte recovery was found to range between 85.81 and 88.21% for NER and 87.09 to 89.44% for NRN, respectively (supporting information, Table S3), at different studied concentrations. The matrix effect was found to be 10.12–14.13% and 9.61–13.07% for NER and NRN, respectively. In addition, the %RSD values of NER, NRN, and IS were found to be within 2%. These results implied the good extraction of NER and NRN in the plasma of rats and negligible influence of the studied matrix.

3.5.6. Stability of the Analytes

The stability test conditions such as bench-top, short-term, long-term, freeze–thaw, and post-processing were tested for NER, NRN, and IS in the rat plasma. The data in Table 3 indicate the stable nature of the analytes in the plasma of rats under the investigated stress conditions, where good extraction recovery and a lack of interference of any plasma components with analytes were observed.

Table 3. Stability data of the method.

Stability Type	Compound	Concentration (ng·mL^{-1})	RE, %	RSD, %
Bench-top at room temperature (up to 6 h)	NER	100 (LQC)	−11.75	2.74
		1000 (HQC)	−13.74	1.44
	NRN	100 (LQC)	−11.37	2.42
		1000 (HQC)	−14.43	1.22
Short term at room temperature (up to 24 h)	NER	100 (LQC)	−11.27	1.47
		1000 (HQC)	−14.45	0.99
	NRN	100 (LQC)	−12.4	1.38
		1000 (HQC)	−14.33	0.57
Long-term at −20 °C (14 days)	NER	100 (LQC)	−11.75	1.64
		1000 (HQC)	−13.31	2.34
	NRN	100 (LQC)	−13.65	1.62
		1000 (HQC)	−14.35	1.04
Freeze–thaw at −20 °C (up to 3rd cycle)	NER	100 (LQC)	−12.7	1.75
		1000 (HQC)	−11.78	1.23
	NRN	100 (LQC)	−12.2	1.85
		1000 (HQC)	−13.5	0.91
Post-preparation at 8 °C	NER	100 (LQC)	−14.7	3.14
		1000 (HQC)	−14.08	1.86
	NRN	100 (LQC)	−11.6	2.77
		1000 (HQC)	−12.87	1.13

RE: Relative error, RSD: Relative standard deviation.

4. Discussion

NER and NRN are a promising combination of modern and natural constituents for the effective management of breast cancer. The former controls breast cancer by its irreversible binding to epidermal growth factor receptors. In contrast, the latter is a derivative aglycone of hydrogenated flavanone origin that inhibits the migration of cancer cells to other body regions, alongside other health benefits that control disease progression. However, to efficiently monitor the therapeutic output of such a combination of drugs, sound bioanalytical LC-MS/MS methods are necessary. The qualitative identification and accurate quantitative determination of analytes in complex biological samples are key benefits of using such techniques. The current UPLC-MS/MS method was newly developed and validated per the regulatory guidelines governing the bioanalysis of drugs in biological fluids. The uncomplicated recovery of analytes from the rat plasma signifies the aptness of the current hyphenated method for the proposed purpose. Combined greenness assessment using contemporary techniques and a high greenness score (score = 90) for the optimized chromatographic condition aptly separated and quantified the analytes and promised environmental sustainability for future use. The method's validation results were satisfactory, with a sensitive linear range of analyte concentrations that engulfed the trueness, precision, selectivity, and good LLOQ values. The final results of all such studies were beneficial to the guidance and reference values that the USFDA and other regulatory bodies have established. No carryover was observed, and the dilution integrity was acceptable. The recovery of analyte greater than 85%, supported with the least matrix effect, vouches for the method's suitability. Finally, the results of the stability study under different stress conditions suggest the acceptable nature of the analytes in the studied matrix.

5. Conclusions

A UPLC-MS/MS method was developed to simultaneously quantify NER and NRN from rat plasma. During the pre-development and optimization phase, the investigations constituted the use of greenness assessment tools such as NEMI and AES, which construed the greenness of the present method. In addition, sensitive LLOQ values were found to befit the method linearity. Further, the validation results from trueness and precision, selectivity, carryover, dilution integrity, recovery and matrix effect, and stability studies matched method intent and were satisfactory. Hence, after assessing the overall study results, the present method conforms to the requirement of simultaneous bioanalytical quantification of the cited analytes for therapeutic drug monitoring in plasmatic samples. The excellent outcome of this work can potentially be linked to evaluating the pharmacokinetic parameters of the drugs in animals and humans.

Supplementary Materials: The following supporting information can be downloaded at: https://www.mdpi.com/article/10.3390/separations10030167/s1, Table S1: Current UPLC-MS/MS method parameter details, Table S2: Details of MS/MS parameters for studied analytes, Table S3: Data showing extraction recovery and matrix effect; Figure S1: Representative MRM chromatograms of blank plasma samples of (A) NER, (B) I.S. and (C) NRN at 80 ng·mL^{-1} concentration; Figure S2: Representative MRM chromatograms of plasma samples spiked with (A) NER, (B) IS, and (C) NRN at 80 ng·mL^{-1} concentration

Author Contributions: A.A. and S.M.A.—study design and funding support; S.S.P.—method validation, data analysis, manuscript writing; M.A., A.B.A. and W.H.A.—data analysis, funding support; M.A.B., R.A.R. and S.N.M.N.U.—experimental support; M.A.A. and M.R.—manuscript writing; S.B.—software support, data analysis, manuscript writing, and language corrections. All authors have read and agreed to the published version of the manuscript.

Funding: The authors extend their appreciation to the Deanship for Research & Innovation and the Ministry of Education in Saudi Arabia for funding this research work through project number: IFP22UQU4310387DSR180.

Institutional Review Board Statement: The animal study protocol was approved by the Institutional Animal Ethic Committee (IAEC) of Roland Institute of Pharmaceutical Sciences (RIPS) (Berhampur, Odisha 760010, India), with the protocol number 146/Chairman IAEC, RIPS, Berhampur (Approval Date: 13 November 2021).

Data Availability Statement: The data are contained within the article or Supplementary Material. The data presented in this study are available in its tables, figures, and Supplementary Materials.

Conflicts of Interest: The authors declare no conflict of interest.

References

1. Łukasiewicz, S.; Czeczelewski, M.; Forma, A.; Baj, J.; Sitarz, R.; StanisÅ‚awek, A. Breast Cancer-Epidemiology, Risk Factors, Classification, Prognostic Markers, and Current Treatment Strategies-An Updated Review. *Cancers* **2021**, *13*, 4287. [CrossRef] [PubMed]
2. Iancu, G.; Serban, D.; Badiu, C.D.; Tanasescu, C.; Tudosie, M.S.; Tudor, C.; Costea, D.O.; Zgura, A.; Iancu, R.; Vasile, D. Tyrosine kinase inhibitors in breast cancer (Review). *Exp. Ther. Med.* **2022**, *23*, 114. [CrossRef]
3. Schlam, I.; Swain, S.M. HER2-positive breast cancer and tyrosine kinase inhibitors: The time is now. *NPJ Breast Cancer* **2021**, *7*, 56. [CrossRef] [PubMed]
4. Burstein, H.J.; Sun, Y.; Dirix, L.Y.; Jiang, Z.; Paridaens, R.; Tan, A.R.; Awada, A.; Ranade, A.; Jiao, S.; Schwartz, G.; et al. Neratinib, an Irreversible ErbB Receptor Tyrosine Kinase Inhibitor, in Patients With Advanced ErbB2-Positive Breast Cancer. *J. Clin. Oncol.* **2010**, *28*, 1301–1307. [CrossRef]
5. Dhillon, S. Neratinib in Early-Stage Breast Cancer: A Profile of Its Use in the EU. *Clin. Drug Investig.* **2019**, *39*, 221–229. [CrossRef]
6. Kourie, H.R.; Chaix, M.; Gombos, A.; Aftimos, P.; Awada, A. Pharmacodynamics, pharmacokinetics and clinical efficacy of neratinib in HER2-positive breast cancer and breast cancer with HER2 mutations. *Expert Opin. Drug Metabol. Toxicol.* **2016**, *12*, 947–957. [CrossRef]
7. Kanth, M.L.; Kamal, B.R. Development and validation of rp-hplc for estimation of neratinib in bulk and tablet dosage form. *Int. J. Pharm. Sci. Drug Res.* **2019**, *11*, 610–614. [CrossRef]

8. Wani, T.A.; Zargar, S.; Ahmad, A. Ultra Performance Liquid Chromatography Tandem Mass Spectrometric Method Development and Validation for Determination of Neratinib in Human Plasma. *S. Afr. J. Chem.* **2015**, *68*, 93–98. [CrossRef]
9. Alrobaian, M.; Panda, S.S.; Afzal, O.; Kazmi, I.; Alossaimi, M.A.; Al-Abbasi, F.A.; Almalki, W.H.; Soni, K.; Alam, O.; Alam, M.N.; et al. Development of a Validated Bioanalytical UPLC-MS/MS Method for Quantification of Neratinib: A Recent Application to Pharmacokinetic Studies in Rat Plasma. *J. Chromatogr. Sci.* **2021**, *60*, 551–558. [CrossRef]
10. Kiesel, B.F.; Parise, R.A.; Wong, A.; Keyvanjah, K.; Jacobs, S.; Beumer, J.H. LC-MS/MS assay for the quantitation of the tyrosine kinase inhibitor neratinib in human plasma. *J. Pharm. Biome. Anal.* **2017**, *134*, 130–136. [CrossRef]
11. Adaway, J.E.; Keevil, B.G. Therapeutic drug monitoring and LC-MS/MS. *J. Chromatogr. B* **2011**, *883*, 33–49. [CrossRef]
12. Saint-Marcoux, F.; Sauvage, F.o.-L.; Marquet, P. Current role of LC-MS in therapeutic drug monitoring. *Anal. Bioanal. Chem.* **2007**, *388*, 1327–1349. [CrossRef] [PubMed]
13. Shipkova, M.; Svinarov, D. LC-MS/MS as a tool for TDM services: Where are we? *Clin. Biochem.* **2016**, *49*, 1009–1023. [CrossRef]
14. Avataneo, V.; D'Avolio, A.; Cusato, J.; Cant, M.; De Nicolo, A. LC-MS application for therapeutic drug monitoring in alternative matrices. *J. Pharm. Biomed. Anal.* **2016**, *166*, 40–51. [CrossRef]
15. Singh, K.; Bhori, M.; Kasu, Y.A.; Bhat, G.; Marar, T. Antioxidants as precision weapons in war against cancer chemotherapy induced toxicity: Exploring the armoury of obscurity. *Saudi Pharm. J.* **2018**, *26*, 177–190. [CrossRef] [PubMed]
16. Ilghami, R.; Barzegari, A.; Mashayekhi, M.R.; Letourneur, D.; Crepin, M.; Pavon-Djavid, G. The conundrum of dietary antioxidants in cancer chemotherapy. *Nut. Rev.* **2020**, *78*, 65–76. [CrossRef]
17. Wang, R.; Wang, J.; Dong, T.; Shen, J.; Gao, X.; Zhou, J. Naringenin has a chemoprotective effect in MDA-MB231 breast cancer cells via inhibition of caspase-3 and 9 activities. *Oncol. Lett.* **2019**, *17*, 1217–1222. [PubMed]
18. Camargo, C.A.; Gomes-Marcondes, M.C.C.; Wutzki, N.C.; Aoyama, H. Naringin Inhibits Tumor Growth and Reduces Interleukin-6 and Tumor Necrosis Factor α Levels in Rats with Walker 256 Carcinosarcoma. *Anticancer Res.* **2012**, *32*, 129. [PubMed]
19. Jha, D.K.; Shah, D.S.; Talele, S.R.; Amin, P.D. Correlation of two validated methods for the quantification of naringenin in its solid dispersion: HPLC and UV spectrophotometric methods. *SN Appl. Sci.* **2020**, *2*, 698. [CrossRef]
20. Musmade, K.P.; Trilok, M.; Dengale, S.J.; Bhat, K.; Reddy, M.S.; Musmade, P.B.; Udupa, N. Development and validation of liquid chromatographic method for estimation of naringin in nanoformulation. *J. Pharm.* **2014**, *2014*, 864901. [CrossRef]
21. Bhandari, R.; Kuhad, A.; Paliwal, J.K.; Kuhad, A. Development of a new, sensitive, and robust analytical and bio-analytical RP-HPLC method for in-vitro and in-vivo quantification of naringenin in polymeric nanocarriers. *J. Anal. Sci. Technol.* **2019**, *10*, 11. [CrossRef]
22. Ma, Y.; Li, P.; Chen, D.; Fang, T.; Li, H.; Su, W. LC/MS/MS quantitation assay for pharmacokinetics of naringenin and double peaks phenomenon in rats plasma. *Int. J. Pharm.* **2006**, *307*, 292–299. [CrossRef] [PubMed]
23. Magdy, M.A.; Farid, N.F.; Anwar, B.H.; Abdelhamid, N.S. Four Greenness evaluations of two chromatographic methods: Application to fluphenazine HCl and nortriptyline HCl pharmaceutical combination in presence of their potential impurities perphenazine and dibenzosuberone. *Chromatographia* **2002**, *85*, 1075–1086. [CrossRef]
24. Gamal, M.; Naguib, I.A.; Panda, D.S.; Abdallah, F.F. Comparative study of four greenness assessment tools for selection of greenest analytical method for assay of hyoscine N-butyl bromide. *Anal. Methods* **2021**, *13*, 369–380. [CrossRef] [PubMed]
25. United States Food and Drug Administration (USFDA) *Bioanalytical Method Validation—Guidance for Industry*; USFDA: Silver Spring, MD, USA, 2018.
26. Chanduluru, H.K.; Sugumaran, A. Assessment of greenness for the determination of voriconazole in reported analytical methods. *RSC Adv.* **2022**, *12*, 6683–6703. [CrossRef]
27. Mohamed, D.; Fouad, M.M. Application of NEMI, Analytical Eco-Scale and GAPI tools for greenness assessment of three developed chromatographic methods for quantification of sulfadiazine and trimethoprim in bovine meat and chicken muscles: Comparison to greenness profile of reported HPLC methods. *Microchem. J.* **2020**, *157*, 104873.
28. Alabbas, A.B.; Alqahtani, S.M.; Panda, S.S.; Alrobaian, M.; Altharawi, A.; Almalki, W.H.; Barkat, M.A.; Rub, R.A.; Rahman, M.; Mir Najib Ullah, S.N.; et al. Development of a validated UPLC-MS/MS method for simultaneous estimation of neratinib and curcumin in human plasma: Application to greenness assessment and routine quantification. *J. Chromatogr. Sci.* **2022**, bmac067. [CrossRef]

Disclaimer/Publisher's Note: The statements, opinions and data contained in all publications are solely those of the individual author(s) and contributor(s) and not of MDPI and/or the editor(s). MDPI and/or the editor(s) disclaim responsibility for any injury to people or property resulting from any ideas, methods, instructions or products referred to in the content.

Article

Determination of Pterostilbene in Pharmaceutical Products Using a New HPLC Method and Its Application to Solubility and Stability Samples

Nazrul Haq [1], Faiyaz Shakeel [1], Mohammed M. Ghoneim [2], Syed Mohammed Basheeruddin Asdaq [2], Prawez Alam [3], Fahad Obaid Aloatibi [4] and Sultan Alshehri [1,*]

1 Department of Pharmaceutics, College of Pharmacy, King Saud University, Riyadh 11451, Saudi Arabia
2 Department of Pharmacy Practice, College of Pharmacy, AlMaarefa University, Ad Diriyah 13713, Saudi Arabia
3 Department of Pharmacognosy, College of Pharmacy, Prince Sattam Bin Abdulaziz University, Al-Kharj 11942, Saudi Arabia
4 Forensic Medical Services Center in Riyadh, Ministry of Health, Riyadh 12746, Saudi Arabia
* Correspondence: salshehri1@ksu.edu.sa

Citation: Haq, N.; Shakeel, F.; Ghoneim, M.M.; Asdaq, S.M.B.; Alam, P.; Aloatibi, F.O.; Alshehri, S. Determination of Pterostilbene in Pharmaceutical Products Using a New HPLC Method and Its Application to Solubility and Stability Samples. *Separations* **2023**, *10*, 178. https://doi.org/10.3390/separations10030178

Academic Editor: Kenichiro Todoroki

Received: 2 February 2023
Revised: 24 February 2023
Accepted: 1 March 2023
Published: 7 March 2023

Copyright: © 2023 by the authors. Licensee MDPI, Basel, Switzerland. This article is an open access article distributed under the terms and conditions of the Creative Commons Attribution (CC BY) license (https://creativecommons.org/licenses/by/4.0/).

Abstract: The quantification of a natural bioactive compound, pterostilbene (PTT), in commercial capsule dosage form, solubility, and stability samples was carried out using a rapid and sensitive high-performance liquid chromatography (HPLC) approach. PTT was quantified on a Nucleodur (150 mm × 4.6 mm) RP C_{18} column with a particle size of 5 µm. Acetonitrile and water (90:10 *v/v*) made up the mobile phase, which was pumped at a flow speed of 1.0 mL/min. At a wavelength of 254 nm, PTT was detected. The developed HPLC approach was linear in 1–75 µg/g range, with a determination coefficient of 0.9995. The developed HPLC approach for PTT estimation was also rapid (R_t = 2.54 min), accurate (%recoveries = 98.10–101.93), precise (%CV = 0.59–1.25), and sensitive (LOD = 2.65 ng/g and LOQ = 7.95 ng/g). The applicability of developed HPLC approach was revealed by determining PTT in commercial capsule dosage form, solubility, and stability samples. The % assay of PTT in marketed capsules was determined to be 99.31%. The solubility of PTT in five different green solvents, including water, propylene glycol, ethanol, polyethylene glycol-400, and Carbitol was found to be 0.0180 mg/g, 1127 mg/g, 710.0 mg/g, 340.0 mg/g, and 571.0 mg/g, respectively. In addition, the precision and accuracy of stability samples were within the acceptable limit, hence PTT was found to be stable in solution. These results suggested that PTT in commercial products, solubility, and stability samples may be routinely determined using the established HPLC method.

Keywords: dosage form; HPLC; pterostilbene; solubility; stability; validation

1. Introduction

As strong antioxidants, natural polyphenols have a key role in regulating a variety of physiological diseases [1,2]. Pterostilbene (PTT) is one of the polyphenolic antioxidants with the chemical structure shown in Figure 1 [3]. Although it can be found in a wide range of plants and fruits, PTT is primarily derived from *Pterocarpus marsupium* [4–6]. It is used to manage diabetes and hypertension in conventional medical care [7]. In the literature, it has also demonstrated for a number of therapeutic efficacies, including antioxidant [8,9], anti-inflammatory [9], anticancer [9,10], antidiabetic [11], cardioprotective [12], and neuroprotective [13] effects, among others.

Due to its wide spectrum of medicinal efficacies, PTT's quality control and standardization in its commercial polyherbal products are crucial. For the qualitative and quantitative detection of PTT in plant extracts, commercial products, commercial polyherbal products, and biological samples, numerous analytical approaches have been reported. These analytical approaches include ultraviolet (UV) spectrometry, high-performance liquid chromatography (HPLC), and ultra-high-performance liquid chromatography (UHPLC) for the determination of PTT [14–20]. PTT has also been identified in biological samples using several HPLC and UHPLC techniques, either alone or in conjunction with other bioactive chemicals [21–24]. To determine PTT in plant extracts, a high-performance thin-layer chromatography (HPTLC) method has also been used [25]. Recently, a greener and sustainable HPTLC approach has also been used by our research group to determine PTT in commercial capsule dosage forms [26].

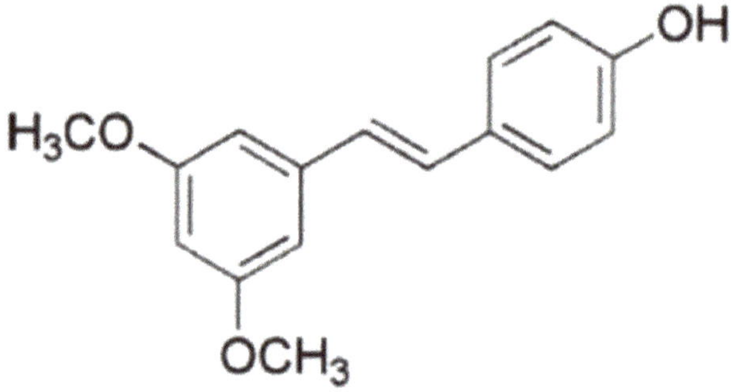

Figure 1. Chemical structure of pterostilbene (PTT).

A thorough literature evaluation suggested a variety of analytical approaches to determine PTT in different kinds of sample matrices. However, the methods in the literature have not been utilized for the measurement of PTT solubility. The solubility of bioactive compounds such as PTT is an important characteristic, and therefore its measurement in a variety of green solvents is important. As a result, simple and cost-effective analytical approaches are still required for its analysis and for further applications. Therefore, the aim of this research was to design and validate a simple, rapid, and cost-effective HPLC approach to determine PTT in commercial capsule dosage forms, solubility, and stability samples. The developed HPLC approach for determining PTT was validated according to the "International Council for Harmonization (ICH)-Q2-R1" protocols [27].

2. Materials and Methods

2.1. Materials

The working standard of PTT was provided by Sigma-Aldrich (St. Louis, MO, USA). Chromatography-grade solvents, including, methanol, acetonitrile, and ethanol were provided by E-Merck (Darmstadt, Germany). High pure water was obtained from Milli-Q® water purifier (Millipore, Lyon, France). All other materials used were of analytical grade. Commercial PTT capsules were bought from the neighborhood pharmacy in Riyadh, Saudi Arabia.

2.2. Instrumentation and Analytical Conditions

PTT was measured at 25 ± 1 °C using a Waters HPLC system 1515 isocratic pump, a 717 automatic sampler, a programmed UV-visible variable wavelength detector, a column oven, a SCL 10AVP system controller, and an inline vacuum degasser. The data were processed and analyzed using Millennium software (version 32, Waters, Milford, MA, USA). PTT was determined using a Nucleodur (150 mm × 4.6 mm) RP C_{18} column with 5 µm-sized particles. The mixture of acetonitrile and water (90:10% v/v) was used as the mobile phase. The mobile phase was flowed with a flow speed of 1.0 mL/min. At a wavelength of 254 nm, PTT was detected. The samples (20 µL) were injected into the system using a Waters autosampler.

2.3. PTT Calibration Curve

An appropriate quantity of PTT was dispensed in a mobile phase to create a PTT stock solution with a 100 µg/g concentration. The required aliquots from the stock solution of

PTT (100 µg/g) were diluted with the mobile phase to create the serial dilutions in the necessary range (1–75 µg/g). Instead of volume/volume, all of the dilutions were prepared on a mass/mass basis. Using the established HPLC method, the chromatographic area for each concentration of PTT was identified. To create the PTT calibration curve, eight different PTT concentrations (1, 5, 10, 15, 20, 25, 50, and 75 µg/g) were plotted against the measured chromatographic area. All the sample preparation and experiments were performed in three replicates (n = 3).

2.4. Analytical Method Development

As the eluent system/mobile phase, various combinations of organic/hydro-organic solvents were investigated for the development of a trustworthy HPLC approach for the detection of PTT in commercial capsules, solubility, and stability samples. The mixtures of methanol–water, ethanol–water, acetonitrile–water, methanol–ethanol, acetonitrile–methanol, acetonitrile–ethanol, methanol–formic acid, ethanol–formic acid, and acetonitrile–formic acid were among the numerous solvents that were investigated. Numerous aspects were taken into account when determining the best solvent or combination of solvents, including the solvents' affordability, the assay's sensitivity, the length of the analysis, the chromatographic parameters and the solvents' compatibility with one another. As a result, different solvents including methanol, ethanol, and acetonitrile were investigated as the mobile phase in both their individual and combined forms with water and formic acid.

2.5. Validation Parameters

Following ICH-Q2-R1 procedures, the developed analytical approach for the measurement of PTT was verified for several parameters [27]. By drawing the linearity plots, the linearity of the developed analytical approach could be investigated in the 1–75 µg/g range. PTT solutions that had just been produced were added to the HPLC apparatus in triplicates (n = 3), and the peak area was estimated. A PTT calibration curve was obtained by plotting PTT concentration vs. peak area.

The peak symmetry, tailing factor (As), capacity factor (k), and theoretical plates number (N) were obtained to examine the system suitability parameters for the developed analytical approach [28,29].

The developed analytical approach's intra-day and inter-day accuracy was estimated using the percent recovery technique. Three replicates (n = 3) were performed on the same day to test intra-day accuracy at three different quality control (QC) levels: low QC (LQC = 10 µg/g), middle QC (MQC = 50 µg/g), and high QC (HQC = 75 µg/g). On three separate days, three replicates (n = 3) of the PTT's LQC, MQC, and HQC levels were used to test inter-day accuracy. The percentage recovery, percentage coefficient of variance (%CV), and standard error were computed for each QC level.

The developed analytical approach's precision was evaluated using intra-day and inter-day variations. On the same day, the same QC levels of PTT (as those used for accuracy) were used to determine the intraday precision. At the same QC levels of the PTT on three consecutive days, inter-day precision was assessed. Both precisions were measured in three replicates (n = 3).

To investigate the impact of intentional chromatographic alterations on PTT analysis, the robustness of the developed analytical approach was evaluated. The PTT MQC (50 µg/g) was selected for the robustness analysis. By adjusting the mobile phase's composition, flow speed, and detecting wavelength, robustness was examined. The initial acetonitrile: water (90:10 *v/v*) mobile phase was adjusted to acetonitrile: water (92:8 *v/v*) and acetonitrile: water (88:12 *v/v*) for the robustness investigation, and the differences in chromatographic response were recorded for each combination of mobile phase. The original flow speed (1 mL/min) was changed to flow rates of 1.1 mL/min and 0.9 mL/min for robustness evaluation by adjusting flow speed, and the variations in chromatographic response were recorded for each set of flow rates. The initial detection wavelength (254 nm) was changed to detection wavelengths of 256 nm and 252 nm for the robustness evaluation by altering the detection wavelength, and the variations in chromatographic response were recorded at each wavelength.

The developed analytical technique's sensitivity was evaluated in terms of the limit of detection (LOD) and limit of quantitation (LOQ), utilizing the standard deviation approach [27]. After the sample was injected into the HPLC system three times (n = 3), the standard deviation of the response was calculated. The LOD and LOQ for PTT were determined using the following equations [27,28]:

$$\text{LOD} = 3.3 \times \frac{\sigma}{S} \quad (1)$$

$$\text{LOQ} = 10 \times \frac{\sigma}{S} \quad (2)$$

where σ is the standard deviation of the response and S is the slope of the calibration curve of PTT.

2.6. Application of Developed HPLC Approach in the Assay of PTT in Commercial Capsules

Ten capsules (each containing an equivalent of 200 mg of PTT) were consumed at random for the test of PTT in commercial capsules, and the average weight was determined. The capsule contents were taken out from the capsule shell and mixed well to obtain the fine powder. The fine powder, with an equivalent to 200 mg of PTT, was dispersed in 100 g of methanol and sonicated for about 15 min. Then, 1 g of this solution was further diluted with methanol to obtain the stock of 100 g. The obtained mixtures of capsules were filtered [26]. The obtained solutions were used for the pharmaceutical assay of PTT in commercial capsules using the developed HPLC approach.

2.7. Application of the Developed HPLC Approach in the Determination of PTT in Solutions

The main purpose of measuring PTT solubility was to enhance the application of the developed method. The solubility of PTT in five different green solvents including water, propylene glycol (PG), ethanol, polyethylene glycol-400 (PEG-400), and Carbitol was determined at 25 °C using a previously reported shake flask method [30]. The excess of PTT was placed into known amounts (10 g) of each green solvent and examined in three replicates (n = 3). The obtained concentrated suspensions were vortexed for about 5 and transferred to a biological shaker for continuous shaking at 100 rpm speed for 72 h [31,32]. The samples were cautiously removed from the shaker once equilibrium had been reached. All the samples were centrifuged at 500 rpm for 30 min. The supernatants from each sample were taken, diluted with mobile phase (wherever required), and subjected to determination of PTT using the developed HPLC approach at a wavelength of 254 nm.

2.8. Application of the Developed HPLC Approach in the Determination of the Stability of PTT in Solutions

The main purpose of determining PTT stability was to enhance the application of developed method. The stability of PTT solution was performed at MQC level (50 μg/g) at two different temperatures, i.e., bench temperature (25 ± 1 °C) and refrigeration temperature (4 ± 0.5 °C). In this work, solution studies were performed; these studies were performed for a short period of time (72 h). The MQC of PTT solution was prepared in mobile phase and stored at 25 ± 1 °C and 4 ± 0.5 °C for about 72 h, and the decomposition of PTT was determined by measuring the rest of PTT after storage.

3. Results and Discussion

3.1. Analytical Method Development

Table 1 provides a summary of the measured chromatographic characteristics and the composition of various eluent systems. The application of methanol and water in various ratios during the analytical method development step led to a subpar chromatographic response of PTT, which exhibited higher As values (As > 1.30) with low N values (<3000).

Additionally, the use of ethanol and water in various ratios caused PTT to have a poor chromatographic response as well as increased As values (As > 1.45) and low N values (<2000). The combination of organic solvents, including acetonitrile and ethanol, acetonitrile and methanol, and methanol and ethanol, was also looked at as an eluent system. With high As values (As > 1.35) and low N values (<3000), the chromatographic response of PTT was once more subpar. We also looked at the binary combinations of organic solvents with formic acid, including methanol: formic acid and ethanol: formic acid. Additionally, the PTT chromatographic response of these binary combinations was subpar, with bigger As values (As > 1.35), and lower N values (<2000).

Table 1. Summary of the eluent systems and measured analytical responses for pterostilbene (PTT) (mean ± SD, n = 3).

Eluent System	As	N	R_t
Methanol: water (50:50 v/v)	1.34 ± 0.02	2478 ± 3.21	2.78 ± 0.04
Methanol: water (90:10 v/v)	1.30 ± 0.03	2771 ± 3.38	2.72 ± 0.05
Ethanol: water (50:50 v/v)	1.47 ± 0.07	1856 ± 2.63	2.68 ± 0.06
Ethanol: water (90:10 v/v)	1.59 ± 0.08	1715 ± 2.54	2.62 ± 0.07
Acetonitrile: water (50:50 v/v)	1.18 ± 0.03	4163 ± 4.22	2.59 ± 0.04
Acetonitrile: water (90:10 v/v)	1.07 ± 0.03	5125 ± 5.84	2.54 ± 0.02
Methanol: ethanol (50:50 v/v)	1.41 ± 0.08	2364 ± 3.11	2.81 ± 0.06
Acetonitrile: methanol (50:50 v/v)	1.37 ± 0.07	2814 ± 3.32	2.74 ± 0.07
Acetonitrile: ethanol (50:50 v/v)	1.44 ± 0.06	2932 ± 3.39	2.71 ± 0.04
Methanol: formic acid (90:10 v/v)	1.38 ± 0.05	1942 ± 2.26	2.86 ± 0.07
Ethanol: formic acid (90:10 v/v)	1.62 ± 0.10	1564 ± 1.97	2.91 ± 0.08
Acetonitrile: formic acid (90:10 v/v)	1.26 ± 0.05	3741 ± 4.51	2.61 ± 0.04

As: tailing factor; N: number of theoretical plates; R_t: retention time.

However, a well-resolved and intact PTT chromatographic peak with good As values and greater N values was shown by the binary mixture of acetonitrile and water in various ratios. The binary mixture of acetonitrile and water (90:10 v/v) gave the best chromatographic response (Figure 2). As a consequence, this mixture was chosen as the final eluent system for measuring PTT, with an acceptable As (1.07) and N (5125), rapid analysis (R_t = 2.54 ± 0.02 min), and a suitable analysis duration (5 min). Therefore, the most trustworthy eluent system for future investigation was a 90:10, volume-to-volume blend of acetonitrile and water.

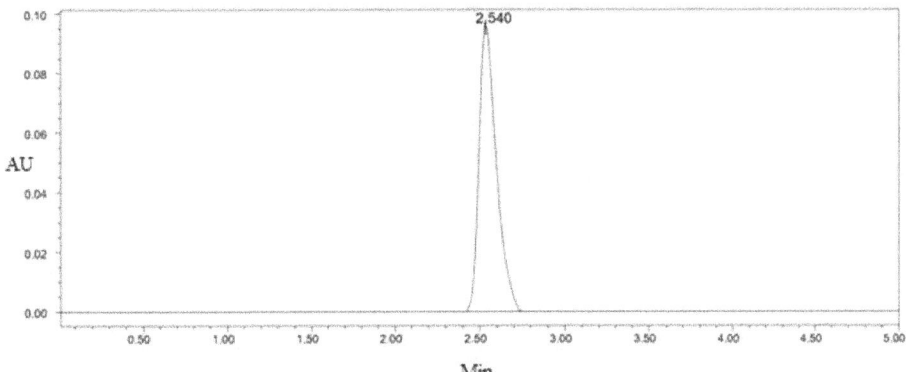

Figure 2. High-performance liquid chromatography (HPLC) chromatogram of PTT (10 µg/g concentration) in solution, produced using a binary eluent system that consisted of acetonitrile and water (90:10 v/v).

3.2. Validation Studies

Several validation parameters for the developed HPLC approach were determined following ICH-Q2-R1 protocols [27]. The linearity graphs were constructed using freshly produced PTT samples (1–75 µg/g). The outcomes of a linear regression analysis of the PTT calibration curve are shown in Table 2. The linear calibration curve for PTT was between 1 and 75 µg/g. According to estimates, the calibration curve's determination coefficient (R^2) and regression coefficient (R) values are 0.9995 and 0.9997, respectively. These outcomes revealed the efficiency of the developed analytical approach for determining PTT.

Table 2. Linear regression analysis for the calibration curve of PTT for the "high-performance liquid chromatography (HPLC)" approach (mean ± SD, n = 3).

Parameters	Values
Linearity range (µg/g)	1–75
Regression equation	y = 9207.2x–4565.5
R^2	0.9995
R	0.9997
Slope ± SD	9207.2 ± 12.13
Intercept ± SD	4565.5 ± 7.41
SE of slope	7.00
SE of intercept	4.27
95% CI of slope	9177.0–9237.3
95% CI of intercept	4547.0–4583.9
LOD (ng/g)	2.65 ± 0.09
LOQ (ng/g)	7.95 ± 0.27

R^2: determination coefficient; R: regression coefficient; SD: standard deviation; SE: standard error; CI: confidence interval; LOD: limit of detection; LOQ: limit of quantification.

The system's appropriateness parameters for the developed analytical approach were determined using the peak symmetry, As, k, and N. The results are shown in Table 3. The developed analytical approach's values for peak symmetry, As, k, and N were found to be satisfactory and acceptable for determining PTT.

Table 3. Optimized chromatographic peak parameters for the resolution of PTT for HPLC approach (mean ± SD, n = 3).

Drug	Peak Symmetry	As	K	N
PTT	1.684 ± 0.11	1.07 ± 0.03	2.78 ± 0.16	5125 ± 5.84

As: tailing factor; k: capacity factor; N: number of theoretical plates.

The percent recovery at LQC, MQC, and HQC was used to determine the intra-day and inter-day accuracy of the established HPLC technique. The results are shown in Table 4. At three different QC levels, the intra-day and inter-day percent recoveries of PTT were found to be 98–102 and 98–101 percent, respectively. According to ICH guidelines, the percent recoveries of analytical method should be within the limit of 100 ± 2% [27]. The percent recoveries of two literature HPLC methods have been reported as 98–99 and 96–100 percent, respectively [17,18]. The percent recoveries of current HPLC method were similar to first reported method [17] and superior to second reported method [18], as per ICH guidelines. High percent recoveries for the established HPLC method for determining PTT point to its accuracy.

The results of the intra-day and inter-day precisions are summarized in Table 5 and are indicated in %CV. For PTT, the intraday variation percent CVs were observed to range from 0.59 to 1.15%. On the contrary, the %CVs for inter-day precision ranged between 0.60 and 1.25 percent. The %CVs of current HPLC method were similar to reported methods [17,18]. Low %CVs in the devised HPLC method for calculating PTT indicated its precision.

Table 4. Intra-day and inter-day accuracy results of PTT for HPLC approach (mean ± SD; n = 3).

Conc. (µg/g)	Intra-Day Accuracy			Inter-Day Accuracy		
	Conc. Found (µg/g) ± SD	Recovery (%)	CV (%)	Conc. Found (µg/g) ± SD	Recovery (%)	CV (%)
10	9.84 ± 0.11	98	1.11	10.13 ± 0.12	101	1.18
50	50.56 ± 0.44	101	0.87	49.05 ± 0.43	98	0.87
75	76.45 ± 0.51	102	0.66	74.44 ± 0.55	99	0.73

Table 5. Intra-day and inter-day precision of PTT for HPLC approach (mean ± SD; n = 3).

Conc. (µg/g)	Intra-Day Precision			Inter-Day Precision		
	Conc. Found (µg/g) ± SD	SE	CV (%)	Conc. Found (µg/g) ± SD	SE	CV (%)
10	10.41 ± 0.12	0.06	1.15	10.32 ± 0.13	0.07	1.25
50	48.96 ± 0.38	0.21	0.77	49.65 ± 0.41	0.23	0.82
75	74.54 ± 0.44	0.25	0.59	76.24 ± 0.46	0.26	0.60

Table 6 contains the results of the robustness assessment for the MQC level of PTT. When evaluating robustness by altering the composition of the mobile phase, the %CV and R_t were discovered to be 0.78–1.18% and 2.53–2.55 min, respectively. The %CV and R_t were found to be 0.43–1.45% and 2.28–2.75 min, respectively, in the scenario of a robustness assessment when the flow speed was changed. The %CV and R_t were calculated to be 1.17–1.55% and 2.55–2.57 min, respectively, in the scenario of a robustness assessment by shifting detecting wavelength. Low CVs and minimal R_t value swings in the devised HPLC method for detecting PTT indicate its robustness.

Table 6. Robustness results of PTT at MQC (50 µg/g) for the HPLC approach (mean ± SD; n = 3).

Parameters	Conc. Found (µg/g) ± SD	CV (%)	R_t ± SD	CV (%)
Mobile phase composition				
(92:8 % v/v)	48.71 ± 0.51	1.04	2.53 ± 0.03	1.18
(88:12 % v/v)	50.68 ± 0.55	1.08	2.55 ± 0.02	0.78
Mobile phase flow rate				
(1.10 mL/min)	50.81 ± 0.61	1.20	2.28 ± 0.01	0.43
(0.90 mL/min)	48.42 ± 0.58	1.19	2.75 ± 0.04	1.45
Detection wavelength (nm)				
252	48.71 ± 0.57	1.17	2.55 ± 0.03	1.17
256	51.11 ± 0.64	1.25	2.57 ± 0.04	1.55

Table 2 lists the findings from evaluating the developed analytical approach's sensitivity in terms of LOD and LOQ. The LOD and LOQ for the developed analytical approach were discovered to be 2.65 ± 0.09 ng/g and 7.95 ± 0.27 ng/g, respectively. These results suggested that the developed analytical approach would have sufficient sensitivity to determine PTT.

The developed HPLC approach for the determination of PTT was compared with reported analytical assays used to determine PTT in solution form. The validation parameters of present HPLC approach compared with reported analytical methods are listed in Table 7. Most of the validation parameters of reported HPLC assays were within the limits of ICH protocol, and hence were similar to the present HPLC approach [17,18]. However, the linearity range, accuracy, precision, LOD, and LOQ values of the HPTLC approaches of PTT analysis in the literature were also found to be inferior to the present HPLC approach [25,26]. Furthermore, the LOD and LOQ values of PTT for the present method were lower than the reported HPLC and HPTLC methods, and were hence found to be more sensitive than the reported HPLC and HPTLC methods. Overall, the newly developed and validated HPTLC approach has been found to be reliable for the determination of PTT.

Table 7. Comparative summary of validation parameters of the present HPLC method with reported methods for the determination of PTT.

Analytical Method	Nature of Sample	Linearity Range	Accuracy (% Recovery)	Precision (% CV)	LOD	LOQ	Ref.
HPLC	Solution	0.02–250 (µg/mL)	98.91–99.59	0.02–0.67	0.006 (µg/mL)	0.019 (µg/mL)	[17]
HPLC	Solution	1–20 (µg/mL)	96.88–100.77	0.20–1.65	0.290 (µg/mL)	0.090 (µg/mL)	[18]
HPTLC	Solution	200–500 (ng/band)	96.67–98.13	0.82–2.12	140 (ng/band)	200 (ng/band)	[25]
Routine HPTLC	Solution	30–400 (ng/band)	90.42–108.82	3.32–3.48	11.1 (ng/band)	33.3 (ng/band)	[26]
Sustainable HPTLC	Solution	10–1600 (ng/band)	98.79–100.94	0.18–0.64	3.51 (ng/band)	10.5 (ng/band)	[26]
HPLC	Solution	1–75 (µg/g)	98.10–101.93	0.59–1.25	2.65 (ng/g)	7.95 (ng/g)	Present work

3.3. Assay of PTT in Marketed Capsules

The developed analytical approach for the PTT assay was shown to be efficient, quick, and sensitive. This approach was therefore used to ascertain PTT in its commercial capsule dosage form. The PTT percentage assay was 99.31% in the commercial capsule dosage form. The PTT percentage in different brands of commercially available capsule dosage forms has been reported as 98.75–98.94% using an HPLC method from the literature [17]. The PTT percentage in the marketed capsule dosage form has been reported as 92.59 and 100.84%, respectively, using routine and sustainable HPTLC methods [26]. The current HPLC method in terms of PTT assay was identical to the reported HPLC and sustainable HPTLC methods [17,26]. However, it was much superior to the reported routine HPTLC method [26]. These findings suggest that the HPLC method would work well for determining PTT in commercially available dosage forms.

3.4. Determination of PTT in Solubility Samples

The potential of the developed HPLC approach was demonstrated by determining the solubility of PTT in five different green solvents, including water, PG, ethanol, PEG-400, and Carbitol, at 25 °C. At 25 °C, the solubility of PTT in water, PG, ethanol, PEG-400, and Carbitol was found to be 0.0180 mg/g, 1127 mg/g, 710.0 mg/g, 340.0 mg/g, and 571.0 mg/g, respectively. Based on these results, PTT was found to be poorly soluble in water, freely soluble in ethanol, PEG-400, and Carbitol, and very soluble in PG [33,34]. Similar solubility characteristics of PTT in water, PG, ethanol, PEG-400, and Carbitol at 25 °C have also been reported in the literature [31]. Hence, the obtained solubility results of PTT were in accordance with those reported in the literature [31]. These results suggested that the developed HPLC approach would be suitable for determining PTT in solubility samples.

3.5. Stability Studies of PTT in Solution

The potential of the developed HPLC approach was also demonstrated by determining the stability of PTT in solution at two different temperatures. The solution of PTT was prepared in mobile phase (acetonitrile: water, 90:10 v/v). The findings of stability evaluations at two different temperatures are included in Table 8. The PTT degradation was measured by determining the rest of PTT concentration after storage. The PTT degradation was very low when held for 72 h at 25 ± 1 °C, and at 4 ± 0.5 °C, when the peak areas of the stored PTT solution were compared to those obtained from a freshly made PTT solution. The precision of PTT in terms of %CV was found to be 1.04–1.07% at two different temperatures. Furthermore, the percent recovery of PTT was found to be 99.84–100.42 percent at two different temperatures. PTT was discovered to be sufficiently stable in solution form at 25 and 4 °C as a result. These findings indicated that PTT stability in solution could be determined using the HPLC method that was established.

Table 8. Stability data of PTT at MCQ level at two different temperatures (mean ± SD; n = 3).

Stability	Nominal Conc. (µg/g)	Conc. Found (µg/g) ± SD	Precision (% CV)	Recovery (%)
Refrigeration (4 °C)	50	49.92 ± 0.52	1.04	99.84
Bench top (25 °C)	50	50.21 ± 0.54 h	1.07	100.42

4. Conclusions

A rapid, sensitive, and economical HPLC approach has been designed and validated for the quantification of PTT in its marketed products, solubility, and stability samples. The developed HPLC approach was validated per ICH-Q2-R1 protocols. The developed analytical approach is rapid, accurate, precise, robust, sensitive, and economical for estimating PTT. The developed HPLC approach was found to be reliable for the determination of PTT in commercial capsule dosage forms, solubility, and stability samples. Based on these findings, it is possible to effectively estimate PTT in a variety of sample matrices using the established HPLC approach. In future, further studies can be carried out to determine PTT in the complex matrices of biological samples, and to accomplish pharmacokinetic assessment of PTT.

Author Contributions: Conceptualization, S.A. and F.S.; methodology, N.H., M.M.G. and S.M.B.A.; software, N.H., F.O.A. and F.S.; validation, S.A. and P.A.; formal analysis, F.O.A. and P.A.; investigation, F.S. and N.H.; resources, S.A.; data curation, P.A.; writing—original draft preparation, F.S.; writing—review and editing, N.H., S.A. and P.A.; visualization, S.A.; supervision, F.S. and S.A.; project administration, F.S. and S.A.; funding acquisition, S.A. and M.M.G. All authors have read and agreed to the published version of the manuscript.

Funding: This research was funded by the Researchers Supporting Project (number RSP2023R146) at King Saud University, Riyadh, Saudi Arabia. This study was also supported via funding from Prince Sattam bin Abdulaziz University project number (PSAU/2023/R/1444). The APC was funded by RSP.

Institutional Review Board Statement: Not applicable.

Informed Consent Statement: Not applicable.

Data Availability Statement: Not applicable.

Acknowledgments: The authors are thankful to the Researchers Supporting Project (number RSP2023R146) at King Saud University, Riyadh, Saudi Arabia for supporting this research. The authors are also thankful to Prince Sattam bin Abdulaziz University for supporting this work via project number (PSAU/2023/R/1444). The authors are also thankful to AlMaarefa University for their generous support.

Conflicts of Interest: The authors declare no conflict of interest.

References

1. Chen, Z.; Farag, M.A.; Zhong, Z.; Zhang, C.; Yang, Y.; Wang, S.; Wang, Y. Multifaceted role of phyto-derived polyphenols in nanodrug delivery systems. *Adv. Drug Deliv. Rev.* **2021**, *176*, E113870. [CrossRef]
2. Di Lorenzo, C.; Colombo, F.; Biella, S.; Stockley, C.; Restani, P. Polyphenols and human health: The role of bioavailability. *Nutrients* **2021**, *13*, E273. [CrossRef] [PubMed]
3. Nagarajan, S.; Mohandas, S.; Ganesan, K.; Xu, B.; Ramkumar, M. New insights into dietary pterostilbene: Sources, metabolism, and health promotion effects. *Molecules* **2022**, *27*, E6316. [CrossRef]
4. Seshadri, T.R. Polyphenols of *Pterocarpus* and *Dalbergia* woods. *Phytochemistry* **1972**, *11*, 881–898. [CrossRef]
5. Mathew, J.; Rao, A. Chemical examination of *Pterocarpus marsupium*. *J. Indian Chem. Soc.* **1984**, *61*, 728–729.
6. Ammulu, M.A.; Viswanath, K.V.; Giduturi, A.K.; Vemuri, P.K.; Mangamuri, U.; Poda, S. Phytoassisted synthesis of magnesium oxide nanoparticles from *Pterocarpus marsupium* rox.b heartwood extract and its biomedical applications. *J. Genet. Eng. Biotechnol.* **2021**, *19*, E21. [CrossRef]

7. Paul, B.; Masih, I.; Deopujari, J.; Charpentier, C. Occurrence of resveratrol and pterostilbene in age-old darakchasava, an ayurvedic medicine from India. *J. Ethnopharmacol.* **1999**, *68*, 71–76. [CrossRef]
8. Waffo Teguo, P.; Fauconneau, B.; Deffieux, G.; Huguet, F.; Vercauteren, J.; Merillon, J.M. Isolation, identification, and antioxidant activity of three stilbene glucosides newly extracted from *Vitis vinifera* cell cultures. *J. Nat. Prod.* **1998**, *61*, 655–657. [CrossRef]
9. Remsberg, C.M.; Yáñez, J.A.; Ohgami, Y.; Vega-Villa, K.R.; Rimando, A.M.; Davies, N.M. Pharmacometrics of pterostilbene: Preclinical pharmacokinetics and metabolism, anticancer, antiinflammatory, antioxidant and analgesic activity. *Phytother. Res.* **2008**, *22*, 169–179. [CrossRef]
10. Chiou, Y.; Tsai, M.; Nagabhushanam, K.; Wang, Y.J.; Wu, C.H.; Ho, C.T.; Pan, M.H. Pterostilbene is more potent than resveratrol in preventing azoxymethane (AOM)-induced colon tumorigenesis via activation of the NF-E2-related factor 2 (Nrf2)-mediated antioxidant signaling pathway. *J. Agric. Food Chem.* **2011**, *59*, 2725–2733. [CrossRef]
11. Pari, L.; Satheesh, A.M. Effect of pterostilbene on hepatic key enzymes of glucose metabolism in streptozotocin- and nicotinamide induced diabetic rats. *Life Sci.* **2006**, *79*, 641–645. [CrossRef] [PubMed]
12. Kosuru, R.; Cai, Y.; Kandula, V.; Yan, D.; Wang, C.; Zheng, H.; Li, Y.; Irwin, M.G.; Singh, S.; Xia, Z. AMPK contributes to cardioprotective effects of pterostilbene against myocardial ischemia-reperfusion injury in diabetic rats by suppressing cardiac oxidative stress and apoptosis. *Cell. Physiol. Biochem.* **2018**, *46*, 1381–1397. [CrossRef]
13. Wang, B.; Liu, H.; Yue, L.; Li, X.; Zhao, L.; Yang, X.; Wang, X.; Yang, Y.; Qu, Y. Neuroprotective effects of pterostilbene against oxidative stress injury: Involvement of nuclear factor erythroid 2-related factor 2 pathway. *Brain Res.* **2016**, *1643*, 70–79. [CrossRef]
14. Mukthinuthalapati, M.A.; Kumar, J.S.P. New derivative and differential spectrophotometric methods for the determination of pterostilbene-an antioxidant. *Pharm. Methods* **2015**, *6*, 143–147.
15. Majeed, M.; Majeed, S.; Jain, R.; Mundkur, L.; Rajalakshmi, H.R.; Lad, P.; Neupane, P. A randomized study to determine the sun protection factor of natural pterostilbene from *Pterocarpus marsupium*. *Cosmetics* **2020**, *7*, E16. [CrossRef]
16. Pezet, R.; Pont, V.; Cuenat, P. Method to determine resveratrol and pterostilbene in grape berries and wines using high performance liquid chromatography and highly sensitive fluorimetric detection. *J. Chromatogr. A* **1994**, *663*, 191–197. [CrossRef]
17. Annapurna, M.M.; Venkatesh, B.; Teja, G.R. Development of a validated stability indicating liquid chromatographic method for the determination of pterostilbene. *Indian J. Pharm. Educ. Res.* **2018**, *52*, S63–S70. [CrossRef]
18. Waszczuk, M.; Bianchi, S.E.; Martiny, S.; Pittol, V.; Lacerda, D.S.; Araujo, A.S.D.S.; Bassani, V.L. Development and validation of a specific-stability indicating liquid chromatography method for quantitative analysis of pterostilbene: Application in food and pharmaceutical products. *Anal. Methods* **2020**, *12*, 4310–4318. [CrossRef]
19. Nikam, K.; Bhusari, S.; Wakte, P. High performance liquid chromatography method validation and forced degradation studies of pterostilbene. *Res. J. Pharm. Technol.* **2022**, *15*, 2969–2975. [CrossRef]
20. Bindu, G.H.; Annapurna, M.M. New stability indicating liquid chromatographic method for the determination of pterostilbene in capsules. *Res. J. Pharm. Technol.* **2018**, *11*, 3851–3856. [CrossRef]
21. Remsberg, C.M.; Yanez, J.A.; Roupe, K.A.; Davies, N.M. High-performance liquid chromatographic analysis of pterostilbene in biological fluids using fluorescence detection. *J. Pharm. Biomed. Anal.* **2007**, *43*, 250–254. [CrossRef]
22. Lin, H.S.; Yue, B.D.; Ho, P.C. Determination of pterostilbene in rat plasma by a simple HPLC-UV method and its application in pre-clinical pharmacokinetic study. *Biomed. Chromatogr.* **2009**, *23*, 1308–1315. [CrossRef]
23. Li, J.; Li, D.; Pan, Y.; Hu, J.H.; Huang, W.; Wang, Z.Z.; Xiao, X.; Wang, Y. Simultaneous determination of ten bioactive constituents of Sanjie Zhentong capsule in rat plasma by ultra-high-performance liquid chromatography tandem mass spectrometry and its application to a pharmacokinetic study. *J. Chromatogr. B* **2017**, *1054*, 20–26. [CrossRef]
24. Sun, J.; Huo, H.; Song, Y.; Zheng, J.; Zhao, Y.; Huang, W.; Wang, Y.; Zhu, J.; Tu, P.; Li, J. Method development and application for multi-component quantification in rats after oral administration of Longxuetongluo capsule by UHPLC-MS/MS. *J. Pharm. Biomed. Anal.* **2018**, *156*, 252–262. [CrossRef]
25. Mallavadhani, U.V.; Sahu, G. Pterostilbene: A highly reliable quality-control marker for the Ayurvedic antidiabrtic plant 'Bijasar'. *Chromatographia* **2003**, *58*, 307–312.
26. Alam, P.; Shakeel, F.; Alam, M.H.; Foudah, A.I.; Faiyazuddin, M.; Alshehri, S. Rapid, sensitive, and sustainable reversed-phase HPTLC method in comparison to the normal-phase HPTLC for the determination of pterostilbene in capsule dosage form. *Processes* **2021**, *9*, E1305. [CrossRef]
27. International Conference on Harmonization (ICH). *Q2 (R1): Validation of Analytical Procedures–Text and Methodology*; International Conference on Harmonization: Geneva, Switzerland, 2005.
28. Haq, N.; Alshehri, S.; Alam, P.; Ghoneim, M.M.; Hasan, Z.; Shakeel, F. Green analytical chemistry approach for the determination of emtricitabine in human plasma, formulations, and solubility study samples. *Sus. Chem. Pharm.* **2022**, *26*, E100648. [CrossRef]
29. Haq, N.; Alanazi, F.K.; Samem-Bekhit, M.M.; Rabea, S.; Alam, P.; Alsarra, I.A.; Shakeel, F. Greenness estimation of chromatographic assay for the determination of anthracycline-based antitumor drug in bacterial ghost matrix of *Salmonella typhimurium*. *Sus. Chem. Pharm.* **2022**, *26*, E100642. [CrossRef]
30. Higuchi, T.; Connors, K.A. Phase-solubility techniques. *Adv. Anal. Chem. Instr.* **1965**, *4*, 117–122.
31. Alqarni, M.H.; Haq, N.; Alam, P.; Abdel-Kader, M.S.; Foudah, A.I.; Shakeel, F. Solubility data, Hansen solubility parameters and thermodynamic behavior of pterostilbene in some pure solvents and different (PEG-400 + water) cosolvent compositions. *J. Mol. Liq.* **2021**, *331*, E115700. [CrossRef]

32. Alanazi, A.; Alshehri, S.; Altamimi, M.; Shakeel, F. Solubility determination and three dimensional Hansen solubility parameters of gefitinib in different organic solvents: Experimental and computational approaches. *J. Mol. Liq.* **2020**, *299*, E112211. [CrossRef]
33. Alshehri, S.; Shakeel, F. Solubility determination, various solubility parameters and solution thermodynamics of sunitinib malate in some cosolvents, water and various (Transcutol + water) mixtures. *J. Mol. Liq.* **2020**, *307*, E112970. [CrossRef]
34. Shakeel, F.; Haq, N.; Alsarra, I.A. Equilibrium solubility determination, Hansen solubility parameters and solution thermodynamics of cabozantinib malate in different monosolvents of pharmaceutical importance. *J. Mol. Liq.* **2021**, *324*, E115146. [CrossRef]

Disclaimer/Publisher's Note: The statements, opinions and data contained in all publications are solely those of the individual author(s) and contributor(s) and not of MDPI and/or the editor(s). MDPI and/or the editor(s) disclaim responsibility for any injury to people or property resulting from any ideas, methods, instructions or products referred to in the content.

Article

Evaluation of Chinese Prickly Ash and Cinnamon to Mitigate Heterocyclic Aromatic Amines in Superheated Steam-Light Wave Roasted Lamb Meat Patties Using QuEChERS Method Coupled with UPLC-MS/MS

Raheel Suleman [1,2], Muawuz Ijaz [3], Huan Liu [1], Alma D. Alarcon-Rojo [4], Zhenyu Wang [1] and Dequan Zhang [1,*]

[1] Institute of Food Science and Technology, Chinese Academy of Agricultural Sciences, Key Laboratory of Agro-Products Processing, Ministry of Agriculture and Rural Affairs, Beijing 100193, China
[2] Department of Food Science and Technology, Faculty of Food Science and Nutrition, Bahauddin Zakariya University, Multan 60000, Pakistan
[3] Department of Animal Sciences, University of Veterinary and Animal Sciences, Jhang Campus, Jhang 35200, Pakistan
[4] Department of Animal Science and Ecology, Autonomous University of Chihuahua, Chihuahua 31453, Mexico
* Correspondence: dequan_zhang0118@126.com; Tel./Fax: +86-10-62818740

Citation: Suleman, R.; Ijaz, M.; Liu, H.; Alarcon-Rojo, A.D.; Wang, Z.; Zhang, D. Evaluation of Chinese Prickly Ash and Cinnamon to Mitigate Heterocyclic Aromatic Amines in Superheated Steam-Light Wave Roasted Lamb Meat Patties Using QuEChERS Method Coupled with UPLC-MS/MS. *Separations* 2023, 10, 323. https://doi.org/10.3390/separations10060323

Academic Editor: Faiyaz Shakeel

Received: 13 May 2023
Revised: 21 May 2023
Accepted: 22 May 2023
Published: 25 May 2023

Copyright: © 2023 by the authors. Licensee MDPI, Basel, Switzerland. This article is an open access article distributed under the terms and conditions of the Creative Commons Attribution (CC BY) license (https://creativecommons.org/licenses/by/4.0/).

Abstract: Chinese prickly ash and cinnamon contain many antioxidants, which scavenge free radicals and can reduce many harmful compounds, such as heterocyclic aromatic amines (HAAs). Modern technologies used for cooking, such as the use of superheated steam roasting, are beneficial in decreasing the development of HAAs. The current study was based on the use of these two spices in roasted lamb patties to mitigate the formation of HAAs in superheated steam roasted patties. Results exhibited significant differences ($p < 0.05$) in the content of both polar and non-polar HAAs as compared to control patties. In cinnamon roasted patties, polar HAAs were reduced from 23.76 to 10.56 ng g^{-1}, and non-polar HAAs were reduced from 21.34 to 15.47 ng g^{-1}. In Chinese prickly ash patties, polar and non-polar HAAs were 43.60 ng g^{-1} and 35.74 ng g^{-1}, respectively. Similarly, cinnamon-treated patties showed a significantly higher ($p < 0.05$) reduction in polar HAAs (23.52 to 12.41 ng g^{-1}) than non-polar (16.08 to 9.51 ng g^{-1}) at concentrations of 0.5–1.5%, respectively, as compared to the control, with 45.81 ng g^{-1} polar and 35.09 ng g^{-1} non-polar HAAs. The polar HAAs tested were PhIP, DMIP, IQx, and 8-MeIQx, while the non-polar were harman and norharman. Both spices and superheated steam controlled HAAs to a significant level in lamb meat patties.

Keywords: lamb meat; heterocyclic aromatic amines; roasted; spices

1. Introduction

Heterocyclic aromatic amine (HAA) formation depends on heat transfer, lipid degradation, oxidation, and cooking methods, such as barbecuing and grilling [1]. Heterocyclic amines can be reduced by the application or induction of antioxidants, as well as the modification of cooking methods [2]. The addition of antioxidants to meat has been considered to be an effective strategy to reduce HAA exposure because of the hypothetical free radical pathway leading to HAA formation [3].

According to Adeyeye [4], antioxidants may trap free radicals, such as intermediates of HAAs, to prevent the formation of HAAs. Among spices, cinnamon and Chinese prickly ash are good sources of antioxidants in meat and meat products in Asian countries [5]. In a lipid peroxidation assay test by Thaipong [6], it was estimated that cinnamon showed more significant activity than anise, ginger, licorice, nutmeg, or vanilla. Chinese prickly ash is also consumed in Central Asian countries, such as China, and is used as an important spice [7]. Antioxidant compounds, such as sanshools and sanshoamides, are very beneficial compounds in this spice, and are responsible for its antioxidant activity [8]. The spices can be a beneficial strategy to control HAAs in cooked meat products [9].

Modern cooking methods can be more helpful in the reduction of HAAs than conventional barbecuing or grilling, which include direct contact of the meat with flame [10]. Modern cooking methods, such as infrared grilling and superheated steam-light wave roasting, offer great potential to reduce HAAs in cooked lamb meat products [11]. One new technology introduced superheated steam-light wave roasting, which is applied by using water vapor to form steam in the oven, and has a higher temperature, which can be helpful in controlling HAAs at a very significant level [12]. The study aimed to analyze the effect of spices and superheated steam-light wave roasting to inhibit HAAs in roasted lamb meat patties.

2. Materials and Methods

2.1. Chemicals and Reagents

The HAA standards 2-amino-9H-pyrido [2,3-b]indole (AαC), 2-amino-3-methyl-9H-pyrido[2,3-b]indole (MeAαC), 1-methyl-9Hpyrido[3,4-b]indole (Harman), 9H-pyrido[3,4-b]indole (Norharman), 2-amino-6-methyldipyrido[1,2-a:3′,2′-d]imidazole (Glu-P-1), 2-amino-1,6-dimethylfuro[3,2-e]imidazo[4,5-b]pyridine (IFP), 3-amino-1,4-dimethyl-5H-pyrido[4,3-b]indole (TrP-P-1), 3-amino-1-methyl-5H-pyrido[4,3-b] indole (TrP-P-2), 2-amino-3-methylimidazo[4,5-f]quinoline (IQ), 2-amino-3,4-dimethylimidazo[4,5-f]quinoline (MeIQ), 2-amino-1-methylimidazo[4,5-b]quinoline (IQ[4,5-b]), 2-amino-1-methyl-6-phenylimidazo[4,5-b]pyridine (PhIP), 2-amino-1-methylimidazo[4,5-f] quinoline (ISO-IQ), 2-amino-1,6-dimethylimidazo[4,5-b]pyridine (DMIP), 2-amino-5-phenylpyridine (Phe-P-1), 2-amino-3-methyl-3H-imidazo [4,5-f]quinoxaline (IQx), 2-amino-3,8-dimethylimidazo[4,5-f]quinoxaline (8-MeIQx), 2-amino-3,4,8-trimethylimidazo[4,5-f]quinoxaline (4,8-DiMeIQx), and 2-amino-3,7,8-trimethylimidazo[4,5-f]quinoxaline (7,8-DiMeIQx) were purchased from Toronto Research Chemicals (Canada). The purity of all standards was greater than 99.9%. The other reagents included for HAA extraction and purification included QuEChERS extraction packets, containing 4 g of magnesium sulfate and 1 g of ammonium acetate. The primary and secondary amine (PSA), endcapped C-18EC extraction column, and MgSO4, together in 15 mL centrifuge tubes, were procured from Agilent Technologies (Santa Clara, CA, USA). DPPH, BHT, ammonium acetate, and acetonitrile were purchased from Sigma-Aldrich Co., Ltd. (St. Louis, MO, USA). Chemicals and reagents were obtained in the packaged form and kept at a suitable required temperature until used for analysis.

2.2. Lamb Meat and Spices

A total of 12 fresh lamb shoulder oyster cut muscles of 8 month old sheep were obtained from the Hongbao sheep meat industry of Bayannur, Inner Mongolia, China. The spices, cinnamon powder and Chinese prickly ash powder, were bought from the local spice market in Beijing, China [5].

2.3. Determination of DPPH Activity of Spices

The antioxidant activity of spices was determined using the DPPH scavenging activity of spices, following the method of Thaipong [13]. Three different concentrations (10, 20, and 30 uL/mL) of Chinese prickly ash and cinnamon were prepared. From each concentration of spices, 1 mL of extract solution was added to 2 mL of freshly prepared DPPH solution (0.1 mM in 95% methanol). This mixture was vortexed for 10 s and then placed in a dark place for 30 min. The absorbance was measured at 517 nm wavelength using the UV-vis spectrophotometer (Persee TU-1810 UV-vis; Persee Instruments Co., Ltd., Beijing, China) at room temperature. Lower absorbance indicated higher radical-scavenging activity. Radical-scavenging activity was calculated as the percentage of DPPH discoloration using the following equation:

$$\text{DPPH-radical-scavenging activity \%} = 100 \times [1 - AE/AD],$$

where AE represents the solution absorbance at 517 nm when 1 mL of each spice solution was mixed with 2 mL of 0.1 mmol·L^{-1} DPPH solution after incubation (30 min) at room temperature, and AD represents the absorbance of 2 mL of 0.1 mmol·L^{-1} DPPH solution with 1 mL Milli-Q H$_2$O. The final unit is % DPPH activity of the sample that has decreased the DPPH content by scavenging the radicals.

2.4. Preparation of Lamb Patties with Spices

After the sheep were slaughtered, their carcasses were cooled at 4 °C for 24 h. Later, the shoulder oyster muscles were removed from the carcasses and kept at −80 °C until they were employed in further experiments. To make the patties, slices of the meat from the lamb oyster muscles were used. It was ensured that visible fat from muscles was properly removed. The 12 muscles were individually ground using a grinder with 5 mm blades to produce ground meat and ensure proper grinding of each muscle. Each muscle weighed 10 g after being ground. Fifty grams of fresh ground lamb meat was taken, and spices (cinnamon and Chinese prickly ash) were added in concentrations of 0.5%, 1%, and 1.5% to the ground meat paste to create each patty, using a 6 cm × 1.5 cm mold to make patties of the same size. The patties with two spices were separately prepared in triplicates. Three patty treatments contained spices, and one without spice was used as control. There were 12 patties in each spice-treated group; therefor, a total of 24 patties were made for each spice (cinnamon and Chinese prickly ash) [5].

2.5. Cooking of Patties

The spiced lamb patties were cooked using a superheated steam-light wave roasting method at 240 °C. The patties were roasted for 17 min, which was needed to reach a core temperature of 72 °C in the oven. The patties were cooked in two batches; one batch for Chinese prickly ash and the other for cinnamon. A digital data logger (Hangzhou Co., Ltd., Hangzhou, China) with a digital thermometer probe was used to observe the internal temperature of each patty. The lamb patties were further cooled at room temperature for one to two hours, and then packed in zip-lock bags and stored at −20 °C until used [11].

2.6. Determination of HAAs in Roasted Patties

HAAs in roasted patties were determined using the method developed by Hsiao, Chen, and Kao [14]. Two grams of ground patty sample were obtained and places in a centrifuge tube. A total of 10 mL of deionized water and one ceramic stone were placed inside the centrifuge tube containing the sample. The tubes were then mixed for 10 min. After adding 10 mL of acetonitrile containing 1% acetic acid, the tube was shaken once more for 10 min. After mixing, 4 g of anhydrous MgSO$_4$ and 1 g of anhydrous C$_2$H$_3$NaO$_2$ were dispersed. After 1 min, the centrifuge tube was spun at a speed of 3200× g of relative centrifugal force for 10 min at a temperature of 4 °C (RCF). A tube containing 900 mg of anhydrous MgSO$_4$, 300 mg of propylsulfonic acid modified silica (PSA), and 300 mg of C18 was used to purify the supernatant after 10 min of centrifugation. The tube was oscillated for 1 min and centrifuged at 3200× g, at 4 °C, for 5 min.

One milliliter of the supernatant was removed after 1 min of centrifugation and nitrogen was used to freeze-dry it. The freeze-dried sample received 0.2 mL of methanol, which was then vortexed. The material was then filtered using a polyvinylidene difluoride (PVDF) membrane filter with a pore size of 0.22 μm. The samples were then analyzed using UPLC-MS/MS (Agilent model 1290). Mass spectrometric analyses were performed on an AB Sciex API 4000™ triple quadrupole mass spectrometer equipped with an electrospray ionization source to look for heterocyclic aromatic amines [15]. Separation was achieved on a Shim-pack GIST C18 (2.1 × 100 mm, 3 μm, 100 Å) at 37 °C. For further dilution, standard stock solutions containing 10 mg in 5 mL methanol were prepared. To establish calibration curves, LODs, and LOQs, stock solutions of standards mixed solutions with final concentrations of 5, 10, 25, 50, 100, 300, and 500 ppb in methanol were created. The mobile phase for analysis at UPLC-MS/MS was composed of (A) 100% HPLC/UPLC-grade acetonitrile and (B) 10 mM ammonium acetate solution (pH 2.9) [16]. For the purpose of achieving equilibrium in the column, a linear gradient profile with 85% A and 15% B was maintained for the first 8 min, changing to 45% A and 55% B after 13 min, and then to 91% A and 9% B after 16 min. For the analysis, a column temperature of 25 °C was maintained with a flow rate of 0.4 mL/min. The injection volume for each sample's analysis was 2 µL. By using UPLC-MS/MS, each sample was examined for 30 min. Table 1 shows the limits of detection and limits of quantification for all HAAs compounds detected through UPLC-MS/MS.

Table 1. Values of limit of detection and limit of quantification of all HAA compounds.

Compound Name	Limit of Detection ng mL^{-1}	Limit of Quantification ng mL^{-1}
DMIP	0.01	0.03
PhIP	0.02	0.06
Norharman	0.01	0.03
Phe-P-1	0.02	0.06
Harman	1.08	3.24
AαC	0.05	0.15
Glu-P-2	0.01	0.03
MeAαC	0.03	0.09
TrP-P-2	0.01	0.03
ISO-IQ	0.14	0.42
Glu-P-1	0.08	0.24
IQ	0.18	0.54
IQ(4,5-b)	0.24	0.72
IQx	0.01	0.03
IFP	0.05	0.15
TrP-P-1	0.01	0.03
MeIQ	0.01	0.03
MeIQx	0.02	0.06
PhIP	0.03	0.09
7,8-DiMeIQx	0.02	0.06
4,8-DiMeIQx	0.03	0.09

The linear range of 0.1–5.00 ng g^{-1} and recovery was from 54.86% to 108.32% for all HAA standards.

Figure 1 shows the peaks of the standards obtained by UPLC-MS/MS. The internal standard was a mixture of 20 HAA standards, and was used for the analysis of HAAs.

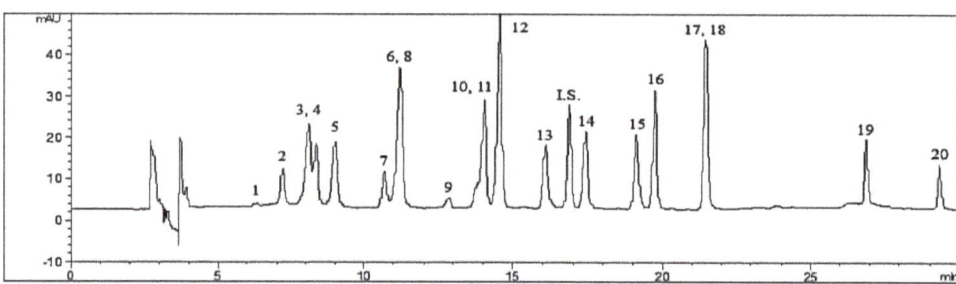

Figure 1. UPLC-MS/MS chromatograms of 20 HAA standards and one internal standard (4,7,8−TriMeIQx) detected by SRM mode. The standard mixture contained 500 ppb of each HA and 4,7,8−TriMeIQx. (1: DMIP; 2: Glu-P-2; 3: Iso-IQ; 4: IQ; 5: IQx; 6: MeIQ; 7: Glu-P−1; 8: 8−MeIQx; 9: IQ[4,5−b]; 10: IFP; 11: 7, 8−DiMeIQx; 12: 4,8−DiMeIQx; 13: Norharman; I.S. (internal standard): 4,7,8−TriMeIQx; 14: Harman; 15: Phe−P−1; 16: Trp-P−2; 17: PhIP; 18: Trp-P−1; 19: AαC; 20: MeAαC.).

3. Results

3.1. DPPH Activity of Cinnamon and Chinese Prickly Ash

Figure 2 shows the DPPH activity of cinnamon and Chinese prickly ash. Both spices showed significant antioxidant activity. Cinnamon showed inhibition rates of 91.96%, 92%, and 91.66% at concentrations of 10, 20, and 30 µL/mL, respectively, while Chinese prickly ash showed inhibition rates of 86.23%, 87.96%, and 89.4% at concentrations of 10, 20, and 30 µL/mL, respectively, which was lower than cinnamon.

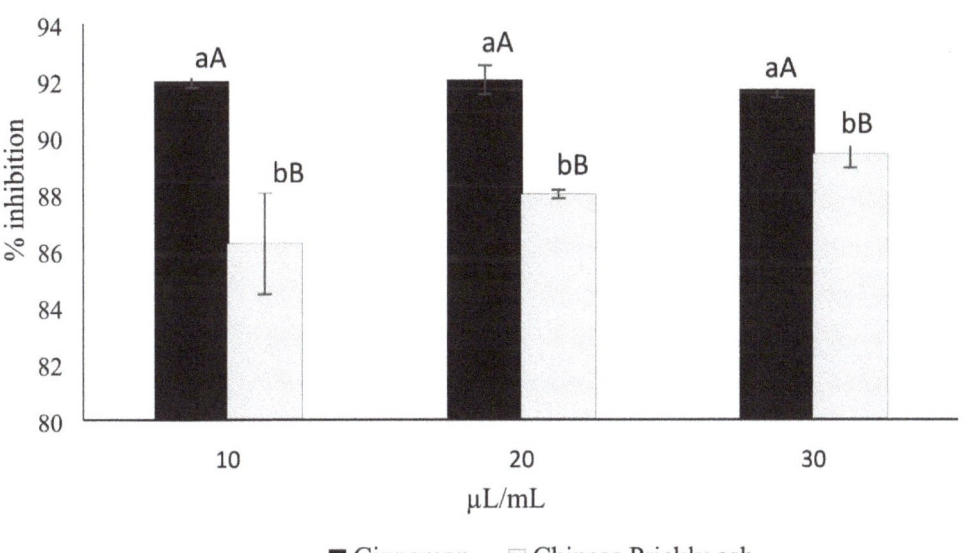

Figure 2. DPPH activity of cinnamon and Chinese prickly ash. Different small (a, b) and capital (A, B) letters show significant differences ($p < 0.05$) among the concentrations of the two spices.

3.2. Quantity of Polar and Non-Polar HAAs in Superheated Steam-Light Wave Roasted Patties and Chinese Prickly Ash

Table 2 lists the findings of the polar HAAs found in the patties treated with Chinese prickly ash. Due to the high antioxidant capacity of Chinese prickly ash, it was shown that its inclusion reduced the production of several polar HAAs [16]. Some polar HAAs, such IQ (4,5-b) and 7,8-DiMeIQx, were not found in detectable amounts. Other polar HAAs, such as IQx, IQ, and 4,8-DiMeIQx, were found in low concentrations; their respective concentrations ranged from 3.25 to 1.91 ng g^{-1}, 0.16 to 0.12 ng g^{-1} (at 1.5% there was no detection of IQ), and 0.78 to 0.38 ng g^{-1}. Results are given as means ± standard errors with superscripts (a, b, c) in columns showing significant differences ($p < 0.05$) within treatments of lamb patties. In beef patties which had Chinese prickly ash also showed a reduction in the polar and the non-polar HAAs [17].

According to the Table 3 findings, the level of non-polar HAAs, such as harman and norharman, which ranged from 8.89 to 6.66 ng g^{-1} and 9.28 to 6.08 ng g^{-1}, respectively, was greater in lamb patties [18]. The concentration of Glu-P-1 ranged from 2.10 to 1.66 ng g^{-1}, and that of Glu-P-2 from 1.29 to 1.12 ng g^{-1}. The lowering of both types of HAA content was observed to be positively impacted by Chinese prickly ash [19]. The Chinese prickly ash-treated patties had lower levels of HAAs than the control samples. According to Figure 3, control patties contained 35.74 ng g^{-1} of non-polar HAAs and 43.60 ng g^{-1} of polar HAAs.

Table 2. Polar HAAs (ng g^{-1}) in the Chinese prickly ash powder-treated patties with superheated steam-light wave roasting.

Treatments	IQx	IQ 4,5-b	IQ	ISO-IQ	DMIP	8-MeIQx	7,8-DiMeIQx	4,8-DiMeIQx	PhIP
Control	6.40 ± 0.14	0.14 ± 0.00 [a]	0.19 ± 0.01 [a]	1.62 ± 0.04 [a]	8.10 ± 0.07 [a]	13.00 ± 0.17 [a]	1.88 ± 0.06 [a]	1.78 ± 0.26 [a]	10.26 ± 0.27 [a]
0.5%	3.25 ± 0.03 [b]	nd [b]	0.16 ± 0.01 [b]	0.19 ± 0.01 [b]	4.03 ± 0.21 [b]	11.00 ± 0.19 [b]	nd [b]	0.78 ± 0.03 [b]	4.26 ± 0.28 [b]
1%	2.77 ± 0.14 [b]	nd [b]	0.12 ± 0.01 [c]	0.17 ± 0.01 [b]	3.44 ± 0.13 [b]	9.80 ± 0.08 [b]	nd [b]	0.72 ± 0.02 [b]	3.60 ± 0.15 [b]
1.5%	1.91 ± 0.01 [c]	nd [b]	nd [d]	nd [c]	2.90 ± 0.23 [b]	3.19 ± 0.31 [c]	nd [b]	0.38 ± 0.01 [b]	2.17 ± 0.05 [c]

Results are given as means ± standard errors with superscripts ([a], [b], [c], [d]) in columns showing significant differences ($p < 0.05$) within treatments of lamb patties.

Table 3. Non-polar HAAs (ng/g) in the Chinese prickly ash powder-treated patties with superheated steam-light wave roasting.

Treatments	Glu-P-1	Norharman	Harman	Glu-P-2
Control	3.38 ± 0.05 [a]	15.28 ± 0.33 [a]	15.56 ± 0.31 [a]	1.52 ± 0.063 [a]
0.5%	2.10 ± 0.02 [b]	9.28 ± 0.08 [b]	8.89 ± 0.36 [b]	1.15 ± 0.00 [b]
1%	1.93 ± 0.03 [b]	7.11 ± 0.20 [c]	7.56 ± 0.17 [bc]	1.12 ± 0.01 [b]
1.5%	1.66 ± 0.15 [b]	6.08 ± 0.29 [c]	6.66 ± 0.34 [c]	1.05 ± 0.017 [b]

Results are given as means ± standard errors with superscripts ([a], [b], [c]) in columns showing significant differences ($p < 0.05$) within treatments of lamb patties.

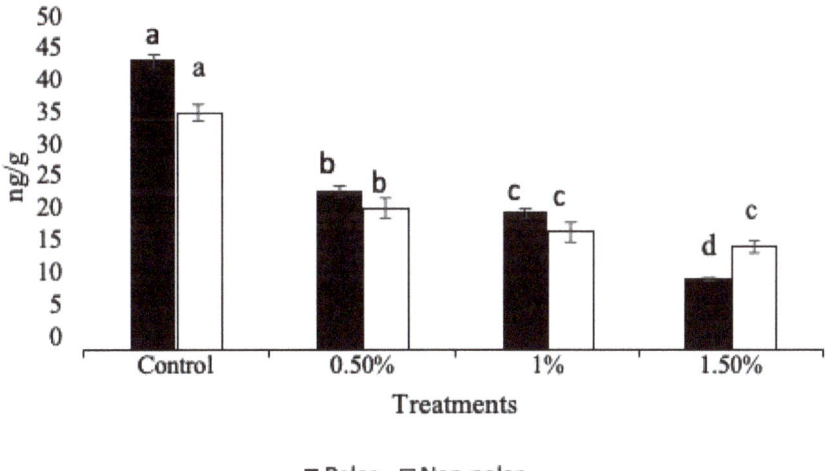

Figure 3. Total polar and non-polar HAAs in Chinese prickly ash lamb patties. Means with different letters (a, b, c, d) denote significant difference ($p < 0.05$) within treatments of lamb patties.

In comparison to the control, the Chinese prickly ash-treated patties had lower concentrations of polar (23.76 ng g^{-1}) and non-polar (21.34 ng g^{-1}) HAAs at 0.5%. At 1%, polar HAA concentrations were marginally lower (20.69 ng g^{-1}), while non-polar HAA concentrations were found to be 17.73 ng g^{-1}. Chinese prickly ash patties contained 10.56 ng g^{-1} of polar HAAs at a concentration of 1.5%, while 15.47 ng g^{-1} of non-polar HAAs were present. The findings demonstrated that the content of the polar HAAs varied significantly between concentrations, as well as when compared to control patties. With the exception of 1.5% concentration, where total non-polar HAAs decreased relative to total polar HAAs, the overall content of polar HAAs was somewhat higher than non-polar HAAs [5].

3.3. Quantity of Polar and Non-Polar HAA Content in Cinnamon and Superheated Steam-Light Wave Roasted Lamb Patties

Table 4 displays the results of the polar HAAs in lamb patties with cinnamon added in various amounts. The polar HAAs were found and contrasted with the no-spice control. The outcomes revealed that all polar HAA levels were greater in the control group [20]. Although there was a significant difference ($p < 0.05$) among the contents seen in the results, they were consistent with the other two spices studied in [21]. The highest values among polar HAAs were observed in 8-MeIQx, DMIP, and PhIP, which ranged from 11.80 to 4.05 ng g^{-1}, 2.93 to 2.24 ng g^{-1}, and 4.47 to 3.16 ng g^{-1}, respectively, and corresponded to treatments 0.5%, 1%, and 1.5% cinnamon powder, respectively. IQ (4,5-b), IQ, ISO-IQ, and 7,8 DiMeIQx content was too low at 1% and 1.5% to be recognized by suppression of cinnamon antioxidants in comparison to the control, while the content was still visible in roasted patties [22].

Table 4. Polar HAAs (ng/g) in the cinnamon powder-treated patties with superheated steam-light wave roasting.

Treatments	IQx	IQ 4,5-b	IQ	ISO-IQ	DMIP	8-MeIQx	7,8-DiMeIQx	4,8-DiMeIQx	PhIP
Control	6.91 ± 0.44 [a]	0.14 ± 0.01 [a]	0.17 ± 0.01 [a]	3.88 ± 0.01 [a]	8.10 ± 0.23 [a]	12.74 ± 0.59 [a]	1.18 ± 0.28 [a]	1.49 ± 0.70 [a]	11.16 ± 0.94 [a]
0.5%	2.91 ± 0.44 [b]	0.12 ± 0.01 [b]	0.13 ± 0.01 [b]	0.17 ± 0.02 [b]	2.93 ± 0.37 [b]	11.80 ± 1.85 [ab]	nd [b]	0.98 ± 0.09 [ab]	4.47 ± 0.17 [b]
1%	2.77 ± 0.49 [b]	nd [c]	0.11 ± 0.01 [b]	0.13 ± 0.01 [b]	2.87 ± 0.83 [b]	9.67 ± 1.04 [b]	nd [b]	0.84 ± 0.02 [ab]	3.26 ± 0.07 [c]
1.5%	2.26 ± 0.21 [b]	nd [c]	nd [c]	nd [c]	2.24 ± 0.03 [b]	4.05 ± 0.87 [c]	nd [b]	0.68 ± 0.02 [b]	3.16 ± 0.14 [c]

Results are given as means ± standard errors with superscripts ([a], [b], [c]) in columns showing significant differences ($p < 0.05$) within treatments of lamb patties.

IQx and 4, 8-DiMeIQx levels were also very low among polar HAAs in the patties with the addition of cinnamon powder, with content reduced from 2.91 to 2.26 ng g^{-1} and 0.98 to 0.68 ng g^{-1}, or from 0.5% to 1.5% concentration, in cinnamon patties. The results of non-polar HAAs, reported in Table 5 below, show that harman and norharman were higher in content than Glu-P-1 and Glu-P-2; harman and norharman levels were 7.65–3.56 ng g^{-1} and 5.95–3.89 ng g^{-1}, respectively, from 0.5% to 1.5%, while the content of Glu-P-1 and Glu-P-2 was detected to be 1.37–1.03 ng g^{-1} and 1.28–1.02 ng g^{-1}, respectively. The results showed that the amount of the polar HAAs was higher generally, and, individually, harman and norharman showed higher content at each concentration, but there was a reduction at all concentrations as compared to content observed in the control.

Table 5. Non-polar HAAs (ng/g) in the cinnamon powder-treated patties with superheated steam-light wave roasting.

Treatments	Glu-P-1	Norharman	Harman	Glu-P-2
Control	3.22 ± 0.12 [a]	14.95 ± 0.90 [a]	15.22 ± 0.58 [a]	1.72 ± 0.13 [a]
0.5%	1.37 ± 0.20 [b]	5.95 ± 0.37 [b]	7.65 ± 0.60 [b]	1.28 ± 0.24 [b]
1%	1.16 ± 0.06 [bc]	4.85 ± 1.01 [bc]	6.89 ± 0.34 [b]	1.19 ± 0.07 [bc]
1.5%	1.03 ± 0.03 [c]	3.89 ± 0.67 [c]	3.56 ± 0.36 [c]	1.02 ± 0.01 [c]

Results are given as means ± standard deviation with different superscripts ([a], [b], [c]) in columns showing significant differences ($p < 0.05$) in the treatments and within treatments of lamb patties.

Cinnamon powder has shown a very potent activity towards the reduction of HAAs after addition in different concentrations [23]. At all concentrations, cinnamon powder had a diminishing effect on the content of HAAs, while the control patties, without cinnamon, had higher contents; the content of polar HAAs was 45.81 ng g^{-1}, while the non-polar HAA content was 35.09 ng g^{-1}, as shown in Figure 4. At 0.5% concentration of cinnamon powder added to the lamb patties, the content of polar HAAs was 23.52 ng g^{-1}, while for non-polar HAAs content was 16.08 ng g^{-1}. At 1%, the content of the polar HAAs was 19.78 ng g^{-1}, while for non-polar HAAs it was 14.04 ng g^{-1}. At 1.5% concentration of cinnamon powder, the content of polar HAAs was 12.41 ng g^{-1}, while for non-polar HAAs it was 9.51 ng g^{-1}. Overall, the results showed that polar HAAs were higher in content both in control as well as in the treated patties with cinnamon powder.

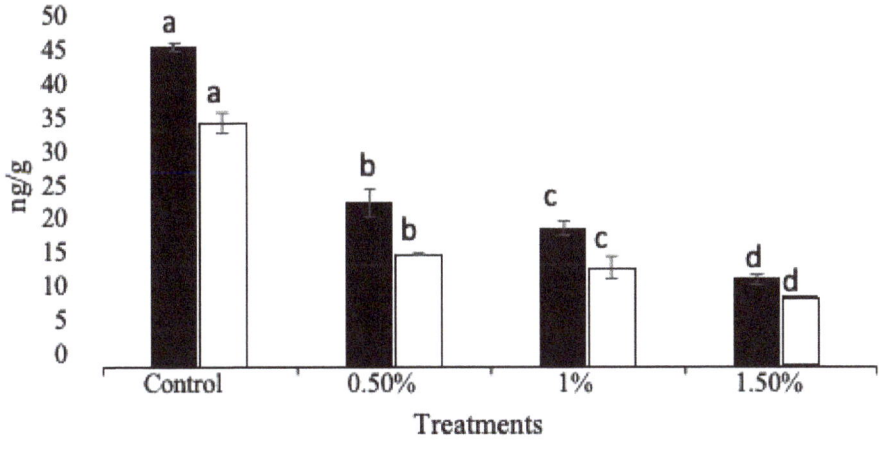

Figure 4. Total polar and non-polar HAAs in the cinnamon lamb patties. Means with different letters (a, b, c, d) denote significant difference ($p < 0.05$) within treatments of lamb patties.

4. Discussion

4.1. DPPH Activity of Cinnamon and Chinese Prickly Ash

Antioxidant compounds present in foodstuffs play a vital role in human life, acting as health-protecting agents. In addition to this role, antioxidants are one of the key additives used in fats and oils [24]. Cinnamon is a very popular spice used in many foods as a flavor additive. It has many other beneficial properties as well, such as anti-inflammatory and antioxidant characteristics [25]. China uses various traditional spices for aroma and taste. One of the most popular spices is Chinese prickly ash, which is added to meat dishes for taste and aroma [7,26]. Based on our results, cinnamon and Chinese prickly ash are both good sources of natural antioxidants, as both showed good or higher levels of inhabitation of DPPH; levels were almost as high as BHT, which is an artificial antioxidant source. Processed meat is cooked at high temperatures, due to which some harmful compounds are produced, such as d-heterocyclic aromatic amines (HAAs) which can cause cancer. The use of cinnamon and Chinese prickly ash in meat products can prevent and block the oxidation of HAAs because of their high antioxidant activity [5].

4.2. Effect of Chinese Prickly Ash Powder and Superheated Steam-Light Wave Roasting on HAAs in Roasted Patties

In China, various traditional spices have become popular around the globe which provide particular taste and aroma in food. Most meat products cooked with these spices are symbolic of Chinese cuisine around the world. One of the most popular spices is Chinese prickly ash (*Zanthoxylum bungeanum*), which is added to meat dishes for taste and aroma [7,26]. Moreover, it is considered to have many important antioxidants and exhibit anti-inflammatory and anti-cancer activity. It contains mainly sanshools and sanshoamides as the main antioxidant compounds [27]. We found that, due to this property, Chinese prickly ash can very efficiently reduce HAAs in roasted lamb meat. The results of the present study are consistent with another study [18], which showed that Chinese prickly ash decreased the content of all polar HAAs. At concentrations of 0.5% to 1% of Chinese prickly ash in roasted beef patties, the contents of PhIP, DMIP, MeIQx, and 4,8-DiMeIQx decreased from 11 to 6.06 ng g^{-1}, 0.42 to 0 ng g^{-1}, 0.69 to 0.32 ng g^{-1}, and 0.25 to 0.24 ng g^{-1}, while in non-polar HAAs, harman content was seen to increase from 0.73 ng g^{-1} to 0.96 ng g^{-1}, and norharman content from 5.12 ng g^{-1} to 6.51 ng g^{-1}, which was opposite to the trend of our results, where we observed a decline of non-polar HAAs at each concentration.

In another study [10], the role of this spice was observed to be very effective in the inhibition of many HAAs in grilled beef. The content of polar HAAs, such as PhIP, was reduced from 1.28 to 0.80 ng g^{-1} by the addition of 0.5–1.5% concentration of Chinese prickly ash, while IQx was only found at 0.5%, with 0.12 ng g^{-1}. MeIQx was reduced from 1.13 to 0.33 ng g^{-1} with a 0.5% to 1.5% concentration of Chinese prickly ash. The content of 4,8-DiMeIQx was reduced from 0.04 ng g^{-1} to 0.02 ng g^{-1} at concentrations of 0.5% to 1.5%. We detected that content of PhIP, IQx, IQ, and 4,8-DiMeIQx was slightly higher than in this study, but the decrease we observed was similar to that observed in this previous study. For non-polar HAAs, the content of harman was reduced from 2.94 to 2.41 ng g^{-1}, while norharman was slightly increased from 16.91 to 17.96 ng g^{-1} at three concentrations, which is quite a contradictory trend compared to our results. In our study, the non-polar HAAs harman and norharman decreased in lamb patties with increasing concentrations of Chinese prickly ash. These results are comparable with those reported by other authors who studied the presence of antioxidant compounds in lamb meat [28].

4.3. Effect of Cinnamon Powder and Superheated Steam-Light Wave Roasting on HAAs in Roasted Patties

Cinnamon (*Cinnamomum cassia* or *Cinnamonum zeylanicum*), which is a popular spice in South Asian and Central Asian countries, is an ordinary spice used on meat products as a flavoring and aromatic agent [29]. Cinnamon has many other beneficial properties, such as anti-inflammatory and antioxidant characteristics [30]. The beneficial bioactive compounds (cinnamaldehyde, cinnamic acid, cinnamyl alcohol, coumarin, and eugenol) in cinnamon make it an important plant spice for medicinal purposes [31]. It is also used to treat many illnesses, such as cancer and diabetes [32]. [33] stated that cinnamon has the potential to inhibit HAAs in beef. Our findings are consistent with a prior study, which found that cinnamon decreased PhIP in descending order, starting with 0.58 ng g^{-1} in beef [33]. There are few publications on DMIP in the literature. Here, we discovered that cinnamon powder in lamb meat considerably ($p < 0.05$) decreased DMIP. In another study [34], it was shown that the addition of cinnamon, at 0.5% concentration, to cooked beef reduced the content of IQ to a lower content than all tested spices. The content of IQ was reduced to the very low level of 0.85 ng g^{-1}, which is very similar to our results.

Similar results were observed in all concentrations, and the contents were reduced to a significant level in lamb meat by the addition of cinnamon. However, in this study PhIP content was increased to 0.43–1.94 ng g^{-1} which is surprising, because we observed a significant decline in our results of PhIP and all polar HAAs. However, the contents of MeIQx and IQx were lower than our values of HAAs at 0.5% cinnamon powder. The MeIQx values were consistent with our results, but there was variation in samples, including kind of meat, cooking method, and temperature variation, as well as cooking method used, between the studies.

5. Conclusions

The spices Chinese prickly ash and cinnamon have great potential as antioxidants, and this property of these spices has an impact on HAA formation. Cinnamon, especially, has great potential to inhibit HAAs, as the antioxidant compounds in it hinders the formation of HAAs. Both kinds of HAAs were reduced by the addition of these spices. Moreover, superheated steam-light wave roasting proved to be a cooking method that was very beneficial in the reduction of HAAs at a significant level as compared to other traditional methods, observed in our previous studies. Therefore, we conclude that the use of cinnamon in meat products can be an important natural source of antioxidants; furthermore, the use of superheated steam-light wave roasting can be a beneficial technology to reduce harmful compounds, which needs further exploration in this area of meat science.

Author Contributions: Conceptualization, R.S.; methodology, R.S.; software, M.I.; validation, R.S. and Z.W.; formal analysis, M.I. and R.S.; investigation, R.S.; resources, R.S.; data curation, M.I.; writing—original draft preparation, R.S.; writing—review and editing, H.L., D.Z. and A.D.A.-R.; visualization, H.L. and D.Z.; supervision, R.S.; project administration, R.S.; funding acquisition, R.S. All authors have read and agreed to the published version of the manuscript.

Funding: This research was funded from the Agricultural Science and Technology Innovation Program (CAAS-ASTIP-2022-IFST-SN2022) and the National Key R&D Program of China (2019YFC1606204).

Data Availability Statement: The datasets generated for this study are available on request to the corresponding author.

Conflicts of Interest: The authors declare no conflict of interest.

References

1. Jian, S.H.; Yeh, P.J.; Wang, C.H.; Chen, H.C.; Chen, S.F. Analysis of heterocyclic amines in meat products by liquid chromatography–Tandem mass spectrometry. *J. Food Drug Anal.* **2019**, *27*, 595–602. [CrossRef]
2. Kang, H.J.; Lee, S.Y.; Kang, J.H.; Kim, J.H.; Kim, H.W.; Oh, D.H.; Jeong, J.W.; Hur, S.J. Main mechanisms for carcinogenic heterocyclic amine reduction in cooked meat by natural materials. *Meat Sci.* **2022**, *183*, 108663. [CrossRef] [PubMed]
3. Khan, I.A.; Luo, J.; Shi, H.; Zou, Y.; Khan, A.; Zhu, Z.; Xu, W.; Wang, D.; Huang, M. Mitigation of heterocyclic amines by phenolic compounds in allspice and perilla frutescens seed extract: The correlation between antioxidant capacities and mitigating activities. *Food Chem.* **2022**, *368*, 130845. [CrossRef]
4. Adeyeye, S.A.; Ashaolu, T.J. Heterocyclic amine formation and mitigation in processed meat and meat products: A Mini-Review. *J. Food Prot.* **2021**, *84*, 1868–1877. [CrossRef] [PubMed]
5. Suleman, R.; Wang, Z.; Hui, T.; Pan, T.; Liu, H.; Zhang, D. Utilization of Asian spices as a mitigation strategy to control heterocyclic aromatic amines in charcoal grilled lamb patties. *J. Food Process. Preserv.* **2019**, *43*, e14182. [CrossRef]

6. Thaipong, K.; Boonprakob, U.; Crosby, K.; Cisneros-Zevallos, L.; Byrne, D.H. Comparison of ABTS, DPPH, FRAP, and ORAC assays for estimating antioxidant activity from guava fruit extracts. *J. Food Compos. Anal.* **2006**, *19*, 669–675. [CrossRef]
7. Murcia, M.A.; Egea, I.; Romojaro, F.; Parras, P.; Jiménez, A.M.; Martínez-Tomé, M. Antioxidant evaluation in dessert spices compared with common food additives. Influence of irradiation procedure. *J. Agric. Food Chem.* **2004**, *52*, 1872–1881. [CrossRef] [PubMed]
8. Zeng, M.; Li, Y.; He, Z.; Qin, F.; Chen, J. Effect of phenolic compounds from spices consumed in China on heterocyclic amine profiles in roast beef patties by UPLC–MS/MS and multivariate analysis. *Meat Sci.* **2016**, *116*, 50–57. [CrossRef]
9. Ji, Y.; Li, S.; Ho, C.T. Chemical composition, sensory properties and application of Sichuan pepper (*Zanthoxylum genus*). *Food Sci. Hum. Wellness* **2019**, *8*, 115–125. [CrossRef]
10. Zeng, M.; Wang, J.; Zhang, M.; Chen, J.; He, Z.; Qin, F.; Xu, Z.; Cao, D.; Chen, J. Inhibitory effects of Sichuan pepper (*Zanthoxylum bungeanum*) and sanshoamide extract on heterocyclic amine formation in grilled ground beef patties. *Food Chem.* **2018**, *239*, 111–118. [CrossRef]
11. Suleman, R.; Hui, T.; Wang, Z.; Liu, H.; Zhang, D. Comparative analysis of charcoal grilling, infrared grilling and superheated steam roasting on the colour, textural quality and heterocyclic aromatic amines of lamb patties. *Int. J. Food Sci. Technol.* **2020**, *55*, 1057–1068. [CrossRef]
12. Suleman, R.; Wang, Z.; Aadil, R.M.; Hui, T.; Hopkins, D.L.; Zhang, D. Effect of cooking on the nutritive quality, sensory properties and safety of lamb meat: Current challenges and future prospects. *Meat Sci.* **2020**, *167*, 108172. [CrossRef] [PubMed]
13. Suwannakam, M.; Noomhorm, A.; Anal, A.K. Influence of combined far-infrared and superheated steam for cooking chicken meat patties. *J. Food Process Eng.* **2014**, *37*, 515–523. [CrossRef]
14. Hsiao, H.Y.; Chen, B.H.; Kao, T.H. Analysis of heterocyclic amines in meat by the quick, easy, cheap, effective, rugged, and safe method coupled with LC-DAD-MS-MS. *J. Agric. Food Chem.* **2017**, *65*, 9360–9368. [CrossRef] [PubMed]
15. Saud, S.; Li, G.; Sun, Y.; Khanm, M.I.; Ur Rehman, A.; Uzzaman, A.; Liu, W.; Ding, C.; Xiao, H.; Wang, Y.; et al. A facile isoelectric focusing of myoglobin and hemoglobin used as markers for screening of chicken meat quality in China. *Electrophoresis* **2019**, *40*, 2767–2774. [CrossRef] [PubMed]
16. Kamankesh, M.; Mollahosseini, A.; Mohammadi, A.; Seidi, S. Haas in grilled meat: Determination using an advanced lab-on-a-chip flat electromembrane extraction coupled with on-line HPLC. *Food Chem.* **2020**, *311*, 125876. [CrossRef] [PubMed]
17. Puangsombat, K.; Jirapakkul, W.; Smith, J.S. Inhibitory activity of Asian spices on heterocyclic amines formation in cooked beef patties. *J. Food Sci.* **2011**, *76*, T174–T180. [CrossRef]
18. Zeng, M.; He, Z.; Zheng, Z.; Qin, F.; Tao, G.; Zhang, S.; Gao, Y.; Chen, J. Effect of six Chinese spices on heterocyclic amine profiles in roast beef patties by ultra-performance liquid chromatography-tandem mass spectrometry and principal component analysis. *J. Agric. Food Chem.* **2014**, *62*, 9908–9915. [CrossRef]
19. Ma, Y.; Fei, X.; Li, J.; Liu, Y.; Wei, A. Effects of location, climate, soil conditions and plant species on levels of potentially toxic elements in Chinese Prickly Ash pericarps from the main cultivation regions in China. *Chemosphere* **2020**, *244*, 125501. [CrossRef]
20. Guo, H.; Wang, Z.; Pan, H.; Li, X.; Chen, L.; Rao, W.; Gao, Y.; Zhang, D. Effects of traditional Chinese cooking methods on formation of heterocyclic aromatic amines in lamb patties. *Food Technol. Biotechnol.* **2014**, *23*, 747–753. [CrossRef]
21. Chiang, C.F.; Liao, P.L.; Hsu, K.C.; Shen, J.Y.; Lin, J.T.; Yang, D.J. Establishment of optimal QuEChERS conditions of various food matrices for rapid measurement of heterocyclic amines in various foods. *Food Chem.* **2022**, *380*, 132184. [CrossRef] [PubMed]
22. Xue, C.; He, Z.; Qin, F.; Chen, J.; Zeng, M. Effects of amides from pungent spices on the free and protein-bound heterocyclic amine profiles of roast beef patties by UPLC–MS/MS and multivariate statistical analysis. *Food Res. Int.* **2020**, *135*, 109299. [CrossRef] [PubMed]
23. Vijayan, V.; Mazumder, A. In vitro inhibition of food borne mutagens induced mutagenicity by cinnamon (*Cinnamomum cassia*) bark extract. *Drug Chem. Toxicol.* **2018**, *41*, 385–393. [CrossRef] [PubMed]
24. Rao, P.V.; Gan, S.H. Cinnamon: A multifaceted *medicinal* plant. *Evid. Based Complement. Altern. Med.* **2014**, *2014*, 642942. [CrossRef]
25. Shobana, S.; Naidu, K.A. Antioxidant activity of selected Indian spices. *Prostaglandins Leukot. Essent. Fat. Acids* **2000**, *62*, 107–110. [CrossRef]
26. Li, J.; Wang, F.; Li, S.; Peng, Z. Effects of pepper (*Zanthoxylum bungeanum* Maxim.) leaf extract on the antioxidant enzyme activities of salted silver carp (*Hypophthalmichthys molitrix*) during processing. *J. Funct. Foods* **2015**, *18*, 1179–1190. [CrossRef]
27. Luo, J.; Ke, J.; Hou, X.; Li, S.; Luo, Q.; Wu, H.; Shen, G.; Zhang, Z. Composition, structure and flavor mechanism of numbing substances in Chinese prickly ash in the genus Zanthoxylum: A review. *Food Chem.* **2022**, *373*, 131454. [CrossRef]
28. Ding, X.; Zhang, D.; Liu, H.; Wang, Z.; Hui, T. Chlorogenic acid and Epicatechin: An efficient inhibitor of heterocyclic amines in charcoal roasted lamb meats. *Food Chem.* **2022**, *368*, 130865. [CrossRef]
29. Hussain, Z.; Li, X.; Zhang, D.; Hou, C.; Ijaz, M.; Bai, Y.; Xiao, X.; Zheng, X. Influence of adding cinnamon bark oil on meat quality of ground lamb during storage at 4 °C. *Meat Sci.* **2021**, *171*, 108269. [CrossRef]
30. Hussain, Z.; Li, X.; Ijaz, M.; Xiao, X.; Hou, C.; Zheng, X.; Ren, C.; Zhang, D. Effect of Chinese cinnamon powder on the quality and storage properties of ground lamb meat during refrigerated storage. *Food Sci. Anim. Resour.* **2020**, *40*, 311. [CrossRef]
31. Yao, Y.; Peng, Z.Q.; Shao, B.; Wan, K.; Shi, J.; Zhang, Y.; Wang, F.; Hui, T. Effects of the antioxidant capacities of 20 spices commonly consumed on the formation of heterocyclic amines in braised sauce beef. *Sci. Agric. Sin.* **2012**, *45*, e4259.

32. Sadeghi, S.; Davoodvandi, A.; Pourhanifeh, M.H.; Sharifi, N.; ArefNezhad, R.; Sahebnasagh, R.; Moghadam, S.A.; Sahebkar, A.; Mirzaei, H. Anti-cancer effects of cinnamon: Insights into its apoptosis effects. *Eur. J. Med. Chem.* **2019**, *178*, 131–140. [CrossRef] [PubMed]
33. Nimkar, M. Inhibition of Heterocyclic Amines in Beef Patties by Spices. Doctoral Dissertation, Kansas State University, Manhattan, KS, USA, 2013.
34. Unal, K.; Karakaya, M.; Oz, F. The effects of different spices and fat types on the formation of heterocyclic aromatic amines in barbecued sucuk. *J. Sci. Food Agric.* **2018**, *98*, 719–725. [CrossRef] [PubMed]

Disclaimer/Publisher's Note: The statements, opinions and data contained in all publications are solely those of the individual author(s) and contributor(s) and not of MDPI and/or the editor(s). MDPI and/or the editor(s) disclaim responsibility for any injury to people or property resulting from any ideas, methods, instructions or products referred to in the content.

Article

Impact and Optimization of the Conditions of Extraction of Phenolic Compounds and Antioxidant Activity of Olive Leaves (*Moroccan picholine*) Using Response Surface Methodology

El Mustapha El Adnany [1], Najat Elhadiri [1], Ayoub Mourjane [2], Mourad Ouhammou [1,*], Nadia Hidar [1], Abderrahim Jaouad [1], Khalid Bitar [3] and Mostafa Mahrouz [1]

[1] Laboratory of Material Sciences and Process Optimization, Faculty of Sciences Semallaia, Cadi Ayyad University, Marrakesh 40000, Morocco; eladnanyelmustapha@gmail.com (E.M.E.A.); elhadiri@uca.ac.ma (N.E.); nadia.hidar@gmail.com (N.H.); jaouad@uca.ac.ma (A.J.); mahrouz@uca.ac.ma (M.M.)

[2] Laboratory of Bioprocesses and Bio Interfaces, Sciences and Technologies Faculty, University Sultan Moulay Slimane, Beni Mella 23000, Morocco; ayoubmourjane@gmail.com

[3] IRis COSmetologie, ZI Al-Massar, Marrakesh 40000, Morocco; bitar.k@ircoslaboratoires.com

* Correspondence: ouhamoumourad@hotmail.com

Abstract: The *Moroccan picholine* tree's leaves contain phenolic compounds that benefit human health. However, the amount and type of these compounds can vary based on factors such as the extraction method and conditions. This study aimed to improve phenolic compounds' extraction while minimising harmful chemicals' use. It has been found that using ethanol as a solvent with ultrasonic extraction is the most effective and environmentally friendly technique. Several parameters, such as the extraction time, solid/solvent ratio, and ethanol concentration as independent variables, were evaluated using a surface response method (RSM) based on the Box–Behnken design (BBD) to optimize the extraction conditions. The experimental data were fitted to a second-order polynomial equation using multiple regression analysis and also examined using the appropriate statistical methods. In optimal conditions, the ultrasonic time, the ratio (solvent/solid) and the concentration (ethanol/water), the content of total polyphenols (TPC), total flavonoids (TFC), and antioxidant activity (by DPPH, ABTS, FRAP) were, respectively, 74.45 ± 1.22 mg EAG/g DM, 17.08 ± 1.85 mg EC/g DM, 83.45 ± 0.89% 82.85 ± 1.52%, and 85.01 ± 2.35%. The identification of phenolic compounds by chromatography coupled with mass spectrum (HPLC-MS) under optimal conditions with two successive extractions showed the presence of hydroxytyrosol, catechin, caffeic acid, vanillin, naringin, oleuropein, quercetin, and kaempferol at high concentrations.

Keywords: olive leaves; extraction; optimization; ultrasound; polyphenols; flavonoids; antioxidant

1. Introduction

The food and pharmaceutical industries are interested in agricultural wastes due to their high content of phenolic bioactive compounds, carbohydrates, oils, and other biochemical molecules [1].

The olive tree is commonly found in the Mediterranean region and is widely spread throughout Morocco, covering 65% of the national tree area. The regions of Fez-Meknes and Marrakech-Safi have the highest concentration of olive-growing areas, covering 54% of the total area and meeting 19% of the demand for edible oils. The olive transformation by-products, including the skin, pulp, pits, and leaves, have caught the attention of the food and pharmaceutical industries due to the presence of phenolic compounds.

Studies have shown that phenolic compounds found in olive tree leaves have beneficial properties, such as antioxidants [2], anticancer, antimicrobial [3], and hypolipidemic activities [4]. However, the amount of phenolic compounds present can vary based on climate, moisture, plant age and variety [5], and extraction methods [6]. Traditional extraction methods, such as maceration and Soxhlet extraction, are slow and yield low amounts of bioactive products [1]. The ultrasonic method, a newer extraction technique, has been developed to efficiently extract organic bioactive compounds from plants [7]. This step is critical in the production of bioactive.

Ultrasonic-assisted extraction (UAE) is considered the greenest extraction process compared to microwave-assisted extraction (MAE), meeting the requirements of the green extraction method [8] as it reduces the temperature, time, and solvent usage [9–11]. This method has been widely used to extract valuable bioactive compounds from various plant materials. One of the food and pharmaceutical industries' dilemmas is improving extraction efficiency while reducing costs, which can be achieved by optimizing extraction conditions [12]. In addition, this technique is usually performed to study some independent factors, requiring more experiments, leading to increased cost and time [13].

Response surface methodology (RSM) is a powerful statistical tool that optimizes complex processes. It has gained popularity for its effectiveness in extracting methods, identifying optimal variable combinations, and simplifying experiment interpretation. This tool has been widely used in various fields [14]. The objective of this study was to examine the effect of certain independent factors of extraction (time, solid/solvent ratio, ethanol (%)) of bioactive compounds (TPC, TFC) and antioxidant activity (DPPH, ABTS, FRAP) with the ultrasonic-assisted extraction (UAE) method using response surface methodology (RSM).

2. Materials and Methods

2.1. Preparation of the Powder

Olive leaves of the Moroccan picholine variety were harvested in Marrakech, Morocco. The leaves were rinsed with water and dried in a ventilated oven (OVEN 19L DRYING AND STERILIZATION DIGITHEAT J.P.SELECTA) with a thickness of 1 cm at 80 °C (according to previous studies [15]) for 5 h (stable weight), then ground using a propeller mill (Mill Grinder For Spices And Professional Coffee, 1 kg). The leaf powder obtained was sieved (digital vibrating laboratory analysis sieve/GKM Siebtechnik GmbH) into four fractions (>125 µm, (125 µm; 50 µm), (50 µm, 25 µm), and <25 µm). The particle size was set at 25–50 µm. The leaf powder was stored at 4 °C in plastic bags.

2.2. Chemicals

The reagents used were pure ethanol, methanol (HPLC grade), Folin–Ciocalteu's, Sodium carbonate (Na_2CO_3), 2,2-diphenyl-1-picrylhydrazyl (DPPH), 2,2′-azino-bis (3-ethylbenzothiazoline-6-sulphonic acid) (ABTS), aluminum trichloride ($AlCl_3$), potassium persulfate ($K_2S_2O_8$), tripyridyltriazine complex (TPTZ), sodium acetate buffer ($C_2H_3 NaO_2$ $3H_2O$ and $C_2H_4O_2$), and hydrochloric acid (HCl).

2.3. Experimental Design and Statistical Analysis

The Box–Behnken design was used to determine the best combination of extraction variables for organic bioactive compounds based on the results of the preliminary single-factor test. Different variables such as extraction time and temperature, sample/solvent ratio, solvent percentage, and pH influence the determination of the content of phytochemicals [16]. Extraction time (min, X1), sample/solvent ratio (g/mL, X2), and ethanol concentration (v/v, X3) were chosen as independent variables, and their coded and uncoded (real) levels of independent variables are shown in Table 1.

Table 1. Coded and actual values for Box–Behnken design (BBD).

Code Symbols	Independent Variables	Level		
		−1	0	+1
X1	Time (min)	30	45	60
X2	Ratio (mL/g)	5	12.5	20
X3	Ethanol (%)	20	60	100

The variation of the response values (Y), with respect to the three variables was fitted into a response surface model and presented in the form of the second-order polynomial equation, is as follows:

$$Yi = \beta 0 + \sum_{i=1}^{k} \beta i Xi + \sum_{i=1}^{k} \beta ii Xi^2 + \sum_{j=1}^{k} \sum_{<i=2}^{k} \beta ij Xi Xj + \varepsilon,$$

where Yi are the experiment responses; β0 represents the theoretical mean value of the response; βi, βj are the coefficients of the linear terms; βii, are the coefficients of the quadratic terms; βij are the coefficients of the interaction terms, and ε the error term.

2.4. Ultrasound-Assisted Extraction of Bioactive Compounds

Extraction with organic solvents has economic and environmental drawbacks. The "green chemistry" concept encourages developing and using less hazardous processes and materials without reducing efficiency [17]. Consequently, solvent extraction of bioactive compounds must be optimized for maximum response using fewer organic solvents. Thus, water was chosen to be studied in combination with ethanol. Ethanol was chosen instead of methanol as the extraction solvent due to the high toxicity of methanol in the human body [18]. Ethanol has the highest affinity for phenolic compounds; therefore, it the first choice for extracting phenolic compounds from fruit and vegetable wastes [19]. We mixed 1 g of olive leaf powder with ethanol/water. The extraction process was performed using a typical ultrasonic apparatus (Heating cleaning bath "Ultrasound HD"-Model 3000866), and the extract was filtered to collect the supernatant. A UV-T80 spectrophotometer was used to analyse the polyphenols, flavonoids, and total antioxidant activity in the samples.

2.5. Total Phenolic Content (TPC) and Total Flavonoid Content (TFC)

The determination of total polyphenols by the method using the Folin–Ciocalteu reagent is described by Singleton et al. [20]. A total of 0.25 mL of leaf extract was mixed with 0.25 mL of Folin–Ciocalteu and 2 mL of distilled water; the mixture was vortexed. After 3 min, 0.25 mL of sodium carbonate (20%) was added; the mixture was stirred and then incubated for 30 min in the dark at room temperature. The absorbance was measured at 750 nm using a UV/VIS spectrophotometer "T80-PG Instruments. The calibration curve for gallic acid was performed, and the results are expressed as mg gallic acid per g dry matter (mg GAE/g DW).

Flavonoid content was determined based on the formation of a flavonoid–aluminum complex that absorbs at 430 nm. The flavonoid assay was performed according to the protocol described by Djeridane et al. (2006) [21]. A total of 1.5 mL of the olive leaf extract was added with 1.5 mL of aluminum trichloride (AlCl3: 2%). After 30 min incubation at room temperature, the absorbance of the reaction mixture was read at 430 nm using a UV/VIS spectrophotometer (T80-PG Instruments). The flavonoid content in the extracts was calculated by reference to a calibration curve established with catechin. Results are expressed as mg catechin equivalent per 1 g dry matter (mg EC/g DW).

2.6. In Vitro Antioxidant Activity

2.6.1. DPPH Radical Reduction Test

For the anti-radical activity of the different extracts of the leaves dried at different temperatures, we used the method based on DPPH (1,1-diphenyl-2-picrylhydrazyl) as a relatively stable radical, according to the protocol described by Abdel Hameed et al. [22]. Briefly, 1 mL of leaf extract was added to 1 mL of DPPH solution (prepared by solubilizing 4 mg of DPPH in 100 mL of ethanol). The mixtures were incubated in the dark for 30 min at room temperature. The decolorization compared to the negative control containing only DPPH solution measured at 517 nm using a UV/visible spectrophotometer type T80. The radical-scavenging activity of DPPH was calculated as follows: %(AA) = ((A517 control − A517 sample)/A517 control) × 100. A517 control is the absorbance of DPPH solution (without sample extract), and A517 sample is the absorbance of the sample with DPPH solution.

2.6.2. ABTS Radical Test

The ABTS•+ radical cation decolorization test also evaluated the anti-radical activity according to the method used by Aadesariya et al. [23]. The ABTS+- radical was generated by the reaction of 7 mM ABTS+ and 2.45 mM potassium persulfate. An equal mixture volume was incubated in the dark for 12–16 h. The ABTS+-solution was diluted with methanol to an absorbance of 0.700 ± 0.02 at 734 nm before use. Then, 1 mL of ABTS+-solution was mixed with 10 µL of leaf extract. The mixture was incubated for 30 min at 30 °C, and the absorbance was measured at 734 nm. The radical scavenging activity was expressed as the percentage of free radical inhibition by the sample and was calculated by the following formula:

ABTS scavenged (%) = [(A734 control − A734 sample)/A517 control] × 100.

A734 control is the absorbance of the control reaction, and A 734 test is the absorbance in the presence of the sample extracts.

2.6.3. Ferric Reducing Antioxidant Power (FRAP) Test

The FRAP (ferric reducing antioxidant power) method is based on the reduction of ferric ions (Fe^{3+}) to ferrous ions (Fe^{2+}). This method evaluates the declining power of compounds at low pH [24]. The ferrous tripyridyltriazine (TPTZ) complex has an intense blue color measured by a spectrophotometer at 593 nm. The FRAP assay was performed according to the protocol of [25]. The FRAP reagent was prepared by mixing 300 mM sodium acetate buffer (3.1 g $C_2H_3NaO_2$ $3H_2O$ and 16 mL $C_2H_4O_2$), pH: 3.6; 10 mM solution of TPTZ in 40 mM HCl; and 20 mM $FeCl_3$ at a ratio of 10:01:01 ($v/v/v$). One hundred microliters of each extract were added to 3 mL of FRAP reagent and 300 µL of H_2O. After incubation at 37 °C for 30 min; the absorbance was measured at 593 nm against the blank [26].

2.7. Model Verification

The extraction conditions were numerically optimized for maximum TPC and TFC content with high antioxidant activities based on regression analysis and 3D surface curves of independent variables. Responses were determined according to the recommended extraction conditions.

2.8. Qualitative and Quantitative Analysis by HPLC-MS

Identification and quantification of phenolic compounds by HPLC-MS were performed according to the method used by Puigventos et al. (2015) [27]. The injection volume of each sample was 10 µL with separation between solvent A (0.1% aqueous formic acid solution) and solvent B (methanol) as follows: 0–3 min, linear gradient from 5 to 25% B; 3–6 min, at 25% B; 6–9 min, from 25 to 37% B; 9–13 min, at 37% B; 13–18 min, from 37 to 54% B; 18–22 min, at 54% B; 22–26 min, from 54 to 95% B; 26–29 min, at 95% B; 29–29. 15 min, back to initial conditions at 5% B; and 29.15 to 36 min, at 5% B. The mobile phase flow rate was 1 mL/min. Ion transfer tube temperature was set at 350° and the full scan MS acquisition mode to m/z 50–1000. The polyphenolic compounds were obtained at 31 min.

2.9. Statistical Analysis

Analysis of variance (ANOVA) and multiple regression analysis were performed to fit the mathematical model using Design Expert 13 software. Significant terms ($p < 0.05$) in the model for each response were found by analysis of variance, and significance was judged by the F statistic calculated from the data. The experimental data were evaluated with various descriptive statistical analyses such as p-value, F-value, sum of squares (SS), degrees of freedom (DF), coefficient variation (CV), the mean sum of squares (MSS), coefficient of determination (R2), and adjusted coefficient determination (Radj2); this is to obtain the statistical significance of the developed quadratic mathematical model.

3. Results and Discussion

3.1. Evaluation and Optimization of Extraction Conditions

This study determined the relationship between response functions and process variables using a three-factor based on the Box–Behnken (BBD) design. The goal was to optimize the extraction conditions for bioactive compounds such as total polyphenols (TPC), total flavonoids (TFC), their corresponding antioxidant activities. Similar scientific studies were used as a reference. [28].

The outcomes of the conducted responses are reported in Table 2. The total polyphenol (TPC) content ranged from 48.69 to 72.98 mg EAG/g DM. The total flavonoid (TFC) content ranged from 10.45 to 16.36 mg EC/g DM. The entire content of TPC and TFC was obtained for trials 13, 14, and 15 under the experimental conditions of X1 = 45 min; X2 = 12.5 mL/g; and X3 = 60%. Regarding DPPH radical scavenging capacity, ABTS and FRAP ranged from 70.95% to 85.69%, 74.67 to 85.41%, and 71.92 to 86.95%, respectively. The highest antioxidant activity was obtained for tests 2 and 13 under X1 = 60 min; X2 = 5 mL/g; X3 = 60% and X1 = 45 min; X2 = 5 mL/g; and X3 = 60%, respectively. Based on these data, the extraction process was optimized to achieve the maximum desirable response.

Table 2. The experimental run from Box–Behnken design (BBD).

N°Exp	Time (min; X_1)	Ratio (mL/g; X_2)	Concentration (%; X_3)	TPC (mg EAG/g DM; Y1)		TFC (mg EC/g DM; Y2)		DPPH (%; Y3)		ABTS (%; Y4)		FRAP (%; Y5)	
				Reel	Predicts	Reel	Predicts	Reel	Predicts	Reel	Predicts	Reel	Predicts
1	30(−1)	5(−1)	60(0)	59.23	58.65	11.12	11.14	76.12	74.97	76.34	76.89	73.65	72.21
2	60(1)	5(−1)	60(0)	66.36	63.65	13.96	14.05	85.69	84.59	85.41	84.05	76.55	79.15
3	30(−1)	20(1)	60(0)	52.98	55.69	11.85	11.77	76.25	77.35	79.69	81.05	82.65	80.05
4	60(1)	20(1)	60(0)	53.31	53.90	11.96	11.95	75.36	76.51	76.23	75.68	73.76	75.20
5	30(−1)	12.5(0)	20(−1)	56.96	56.02	11.08	10.88	77.36	76.46	75.66	74.58	72.36	74.92
6	60(1)	12.5(0)	20(−1)	55.25	56.43	12.15	11.88	78.85	81.84	79.84	81.22	76.02	74.53
7	30(−1)	12.5(0)	100(1)	48.69	47.51	10.69	10.96	70.95	80.99	83.46	82.08	74.12	75.61
8	60(1)	12.5(0)	100(1)	49.36	50.31	12.85	13.05	78.36	79.26	77.24	77.77	80.65	78.09
9	45(0)	5(−1)	20(−1)	60.26	61.79	10.96	11.15	75.28	77.34	76.08	76.06	75.14	74.03
10	45(0)	20(1)	20(−1)	56.45	54.69	10.45	10.74	76.52	76.33	74.67	73.84	71.92	71.96
11	45(0)	5(−1)	100(1)	51.96	50.31	12.39	12.10	77.39	79.58	76.87	77.70	72.18	72.14
12	45(0)	20(1)	100(1)	49.65	48.12	11.23	11.04	74.96	72.90	75.69	75.71	76.98	78.10
13	45(0)	12.5(0)	60(0)	72.98	72.41	15.99	16.34	82.25	81.65	82.96	82.65	86.95	86.28
14	45(0)	12.5(0)	60(0)	72.26	72.41	16.98	16.34	81.12	81.65	83.23	82.65	86.45	86.28
15	45(0)	12.5(0)	60(0)	71.99	72.41	16.06	16.34	81.59	76.99	81.75	82.65	85.45	86.28

Pearson's test showed a strong positive correlation, r = 0.8- 0.85, between TPC and TFC and between TFC and FRAP. This implies that these answers evolve proportionally. The positive average correlations r = 0.5–0.75 appear between the other answers.

3.2. Fitting the Model and Analysis of Variance

The extraction process was optimised by applying the second-order polynomial model fit. The results are presented in Table 3. The model shows a high significance level and a good fit, with the experimental data of TPC and TFC contents showing less variation around the mean (R^2 values 0.969 and 0.991), respectively.

Table 3. Experimental design of the surface response and statistical table of results.

Source	Sum of Squares	Estimation of Coefficients	Degree of Freedom	Medium Square	Value F	Value p	Remarks
			TPC				
Model	990.43		9	11005	17.53	0.0028	significant
Intercept, X_0		72.41 *					
Linear							
X_1	5.15	0.8025	1	5.15	0.8209	0.4065	
X_2	80.77	−3.18 *	1	80.77	12.87	0.0157	
X_3	107.02	−3.66 *	1	107.02	17.05	0.0091	
Interaction							
$X_1 X_2$	11.56	−1.70	1	11.56	1.84	0.2328	
$X_1 X_3$	1.42	0.5950	1	1.42	0.2256	0.6548	
$X_2 X_3$	0.5625	0.3750	1	0.5625	0.0896	0.7767	
Quadratic							
X^2_1	249.94	−8.23 *	1	249.94	39.83	0.0015	
X^2_2	142.51	−6.21 *	1	142.51	22.71	0.0050	
X^2_3	498.34	−11.62 *	1	498.34	79.40	0.0003	
Residual	31.38		5	6.28			
Lack of fit	30.86		3	10.29	39.27	0.0249	significant
Error	0.5238		2	0.2619			
Total	1021.81		14				
Accuracy Adequacy	12.17						
CV%	4.28						
R2	0.969						
R2Ajust	0.91						
Average	58.51						
			TFC				
Model	61.99		9	6.89	31.55	0.0007	significant
Intercept, X_0		16.34 *					
Linear							
X_1	4.77	0.7725 *	1	4.77	21.87	0.0055	
X_2	1.08	−0.3675	1	1.08	4.95	0.0767	
X_3	0.7938	0.3150	1	0.7938	3.64	0.1148	
Interaction							
$X_1 X_2$	1.86	−0.6825 *	1	1.86	8.54	0.0330	
$X_1 X_3$	0.2970	0.2725	1	0.2970	1.36	0.2960	
$X_2 X_3$	0.1056	−0.1625	1	0.1056	0.4838	0.5177	
Quadratic							
X^2_1	12.54	−1.84 *	1	12.54	57.44	0.0006	
X^2_2	19.16	−2.28 *	1	19.16	87.76	0.0002	
X^2_3	29.11	−2.81 *	1	29.11	133.35	<0.0001	
Residual	1.09		5	0.2183			
Lack of fit	0.4811		3	0.1604	0.5253	0.7074	Not significant
Error	0.6105		2	0.3052			
Total	63.08		14				
Accuracy Adequacy	14.69						
CV%	3.69						
R2	0.98						
R2Ajust	0.95						
Average	12.65						
			DPPH				
Model	165.56		9	18.40	5.19	0.0423	significant
Intercept, X_0		81.65 *					
Linear							
X_1	38.63	2.20 *	1	38.63	10.89	0.0215	
X_2	16.22	−1.42	1	16.22	4.57	0.0855	
X_3	5.04	−0.7937	1	5.04	1.42	0.2867	
Interaction							
$X_1 X_2$	27.35	−2.61 *	1	27.35	7.71	0.0390	
$X_1 X_3$	8.76	1.48	1	8.76	2.47	0.1768	
$X_2 X_3$	3.37	−0.9175	1	3.37	0.9493	0.3747	
Quadratic							
X^2_1	8.06	−1.48	1	8.06	2.27	0.1920	
X^2_2	12.24	−1.82	1	12.24	3.45	0.1224	

Table 3. Cont.

Source	Sum of Squares	Estimation of Coefficients	Degree of Freedom	Medium Square	Value F	Value p	Remarks
X^2_3	53.19	−3.80 *	1	53.19	14.99	0.0117	
Residual	17.74		5	3.55			
Lack of fit	17.09		3	5.70	17.68	0.0540	Not significant
Error	0.6445		2	0.3222			
Total	183.30		14				
Accuracy Adequacy	8.2473						
CV%	2.42						
R2	0.9032						
R2Ajust	0.73						
Average	77.87						
ABTS							
Model	163.76		9	18.20	8.05	0.0168	significant
Intercept, X_0		82.65 *					
Linear							
X_1	1.59	0.4462	1	1.59	0.7045	0.4395	
X_2	8.86	−1.05	1	8.86	3.92	0.1046	
X_3	6.14	0.8762	1	6.14	2.72	0.1602	
Interaction							
$X_1 X_2$	39.25	−3.13 *	1	39.25	17.36	0.0088	
$X_1 X_3$	27.04	−2.60 *	1	27.04	11.96	0.0181	
$X_2 X_3$	0.0132	0.0575	1	0.0132	0.0058	0.9420	
Quadratic							
X^2_1	0.0000	−0.0033	1	0.0000	0.0000	0.9968	
X^2_2	38.42	−3.23 *	1	38.42	16.99	0.0092	
X^2_3	47.68	−3.59 *	1	47.68	21.08	0.0059	
Residual	11.31		5	2.26			
Lack of fit	10.06		3	3.35	5.40	0.1602	Not significant
Error	1.24		2	0.6212			
Total	175.07		14				
Accuracy Adequacy	8.31						
CV%	1.90						
R2	0.94						
R2Ajust	0.82						
Average	79.01						
FRAP							
Model	364.85		9	40.54	5.21	0.0418	significant
Intercept, X_0		86.28 *					
Linear							
X_1	2.20	0.5250	1	2.20	0.2836	0.6171	
X_2	7.59	0.9737	1	7.59	0.9757	0.3686	
X_3	9.01	1.06	1	9.01	1.16	0.3309	
Interaction		0.5250					
$X_1 X_2$	34.75	−2.95	1	31.75	4.47	0.0881	
$X_1 X_3$	2.06	0.7175	1	2.06	0.2649	0.6287	
$X_2 X_3$	16.08	2.00	1	16.08	2.07	0.2099	
Quadratic							
X^2_1	98.21	−3.95 *	1	57.58	7.41	0.0417	
X^2_2	275.63	−5.68 *	1	119.19	15.33	0.0112	
X^2_3	101.96	−6.55 *	1	158.25	20.36	0.0063	
Residual	57.58		5	7.77			
Lack of fit	119.19		3	12.57	21.55	0.0447	significant
Error	158.25		2	0.5833			
Total	38.87		14				
Accuracy Adequacy	6.29						
CV%	3.59						
R2	0.90						
R2Ajust	0.73						
Average	77.79						

* Significant ($p < 0.05$).

The antioxidant activity (DPPH, ABTS, and FRAP) shows that the model is significant, and the polynomial equation fit using the coefficient of determination (R2) 0.90, 0.93, and 0.90, respectively. The regression coefficients for the dependent variables were obtained by multiple linear regressions, as shown in Table 3.

- The linear effect of extraction time (X_1) was significant for TFC and DPPH;
- The solvent/solid ratio (X_2) was significant for TPC;
- The concentration (X_3) was significant for the TPC;
- The quadratic effect of solvent concentration (X_3) and ratio (X_2) was significant for all responses except DPPH;
- The $X_1 X_2$ interaction effect also significantly impacted TFC, DPPH, and ABTS;
- The $X_1 X_3$ interaction was significant for ABTS.

The ANOVA result for each variable response indicates that at least one of the model parameters can explain the experimental variation of the response variables (Table 3).

The corresponding variables would be more significant if the F-value becomes larger and the *p*-value becomes smaller [29]. The *p*-value < 0.05 showed that the model terms were significant. In terms of coefficients of variation (CV), the models recorded a CV for CPT, CFT, DPPH, ABTS, and FRAP of 4.24%, 3.69%, 2.42%, 1.9%, and 3.59%, respectively.

Generally, the acceptable coefficient of variation (CV) value should be less than 20%. The diagnostic diagram as the predicted versus actual values (Figure 1) evaluates the relationship and model satisfaction between the experimental and predicted values obtained from the developed models. From Figure 1, it is observed that the data points are located near the straight line, which means a high correlation between experimental and predicted data obtained for TFC and a medium correlation between experimental and predicted data of TPC and antioxidant activity (DPPH, ABTS, and FRAP) from the models.

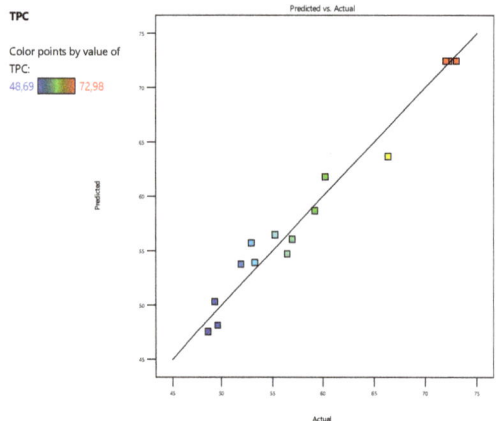
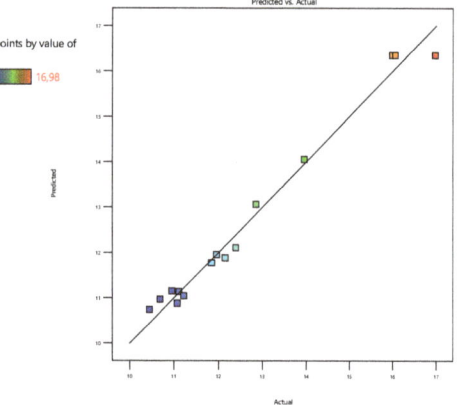

Figure 1. *Cont.*

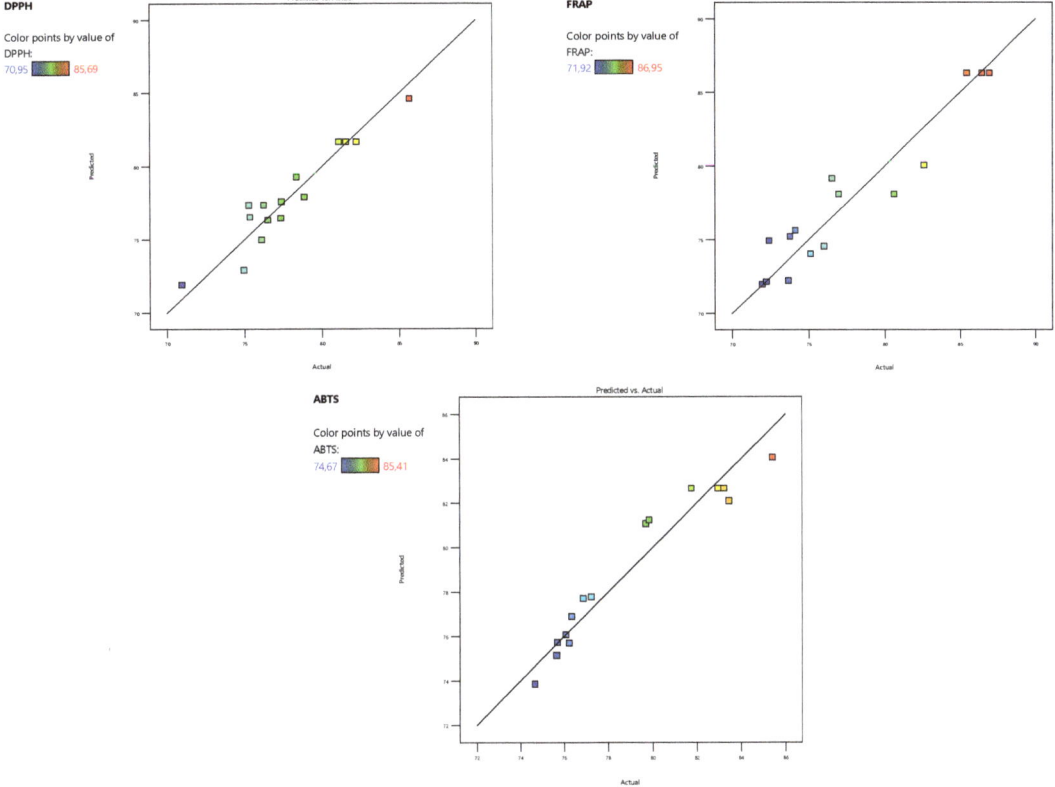

Figure 1. Diagnosis between experimental and predicted values for TPC, TFC, DPPH, ABTS, and FRAP.

3.3. Development of Second Order Polynomial Models

A statistical analysis was conducted using a second-order polynomial equation with interaction terms to model the connection between three process variables and the efficiency of ultrasonic extraction. This model can be utilized to anticipate the efficiency of ultrasonic extraction for varying combinations of process variables.

Five models were developed from this study to have the ultrasound extraction efficiency of TPC, TFC, DPPH, ABTS, and FRAP from olive leaves. The second-order equations of the responses in terms of coded factors are given below:

$$Y_{TPC} = 72.41 - 3.18X_2 - 3.66X_3 - 8.23X^2_{11} - 6.21X^2_2 - 11.62X^2_{33}$$

$$Y_{TFC} = 16.34 + 0.77X_1 - 0.68X_1X_2 - 1.84X^2_{11} - 2.28X^2_{22} - 2.81X^2_{33}$$

$$Y_{DPPH} = 81.65 + 2.2X_1 - 2.61X_1X_2 - 3.8X^2_3$$

$$Y_{ABTS} = 82.65 - 3.13X_1X_2 - 2.60X_1X_3 - 3.23X^2_2 - 3.59X^2_{33}$$

$$Y_{FRAP} = 86.28 - 3.95X^2_1 - 5.68X^2_2 - 6.55X_3^2.$$

3.4. Effect of Process Variables

This study analyzed the impact of process variables on the extraction of bioactive compounds (TPC, TFC, and free radical antioxidants) from olive leaves using a two-level Box–Behnken design with three factors: extraction time, solid/solvent ratio, and ethanol concentration. The results were presented using a three-dimensional response surface, demonstrating the relationship between the independent and dependent variables. By holding two factors constant and varying the third, the response surface curves display the main effects and interactions of the independent variables on/with the dependent variables [1]. These graphs provide insight into how the different variables affect the extraction process.

3.4.1. Effect of Extraction Time

Based on the results obtained, a longer contact time between the sample and solvent leads to a higher transfer rate of bioactive compounds, ultimately resulting in better extraction efficiency. Figure 2 illustrates that the extraction time of 30–45 min was particularly impactful for flavonoids and DPPH. The linear effect of extraction time was significant ($p < 0.05$) for these two variables but not for the other independent variables. This effect is likely due to the extended time the plant matrix is exposed to the solvent, improving the solubility of the leaves' constituents. Essentially, the duration of the solvent exposure facilitates the migration of chemical compounds into the solution [30].

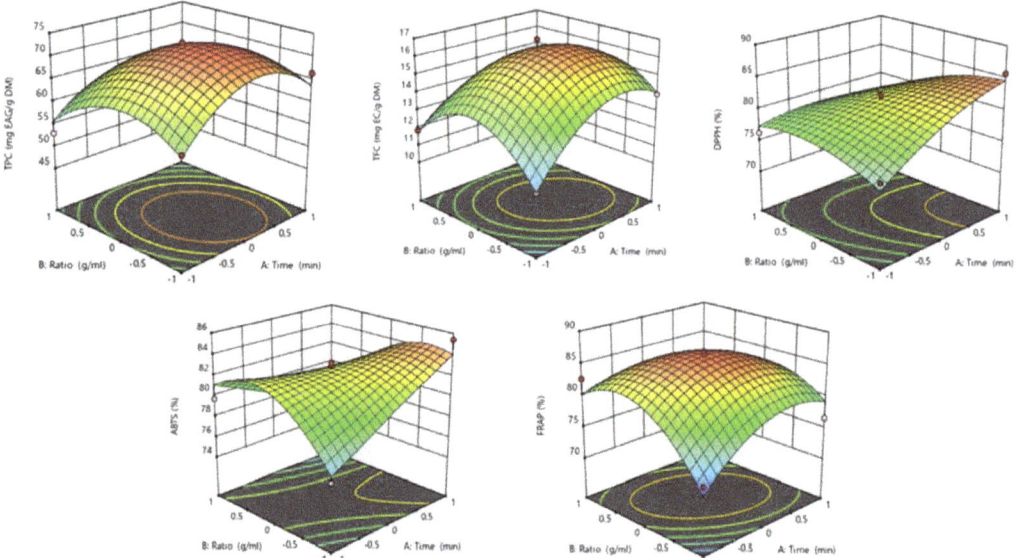

Figure 2. Response surface plot showing the variation of responses as a function of extraction time.

Studies have shown that the longer the extraction time, the higher the content of polyphenols and flavonoids. However, it is essential to note that excessively long extraction times can lead to the degradation of the bioactive compounds, as indicated in [31]. On the other hand, some studies found that extraction time was not a significant factor in the ultrasound-assisted extraction of phenolic compounds [32,33]. In other studies, such as those on Genipap berry pulp, blueberries, and carob pulp, the extraction time for bioactive compounds using ultrasonic-assisted extraction (UAE) was around 49, 50, and 57 min, respectively [34–36].

3.4.2. Effect of Solid–Liquid Ratio

The amount of solvent used in organic bioactive compounds and antioxidant extraction is a crucial factor. When evaluating its impact on the extraction of phenolic compounds, the results, presented in Figure 3, indicate that increasing the solid-liquid ratio up to 12.5 mL/g resulted in an increase in polyphenol and flavonoid content and antioxidant activity. This indicates that the volume of solvent used plays a significant role in achieving good infusion and easy release of bioactive compounds into the surrounding environment [37]. However, according to Zakaria Fazila (2021) [38], a high solvent-to-solid ratio (between 10–30 mL/g) can lead to a decrease in phenolic compound content. The most effective ratio for maximum phenolic compound extraction was found to be 10 mL/g.

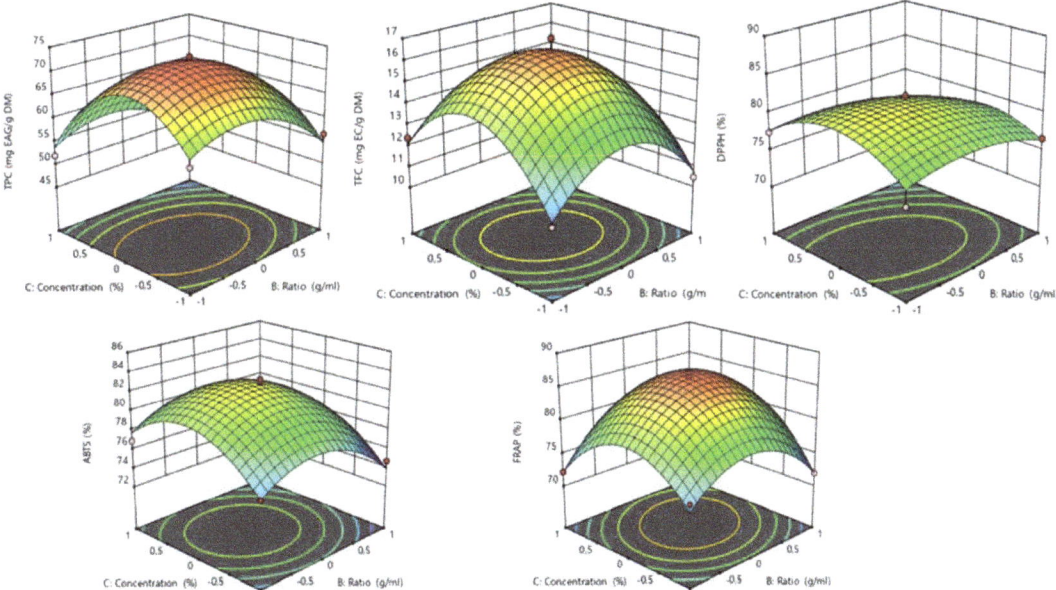

Figure 3. Response surface plot showing the variation of responses as a function of solid/solvent ratio.

This research found that once the solid–liquid ratio surpasses 12.5 mL/g, the solution becomes oversaturated with solute. This can lead to a reduction in the rate of mass transfer and a hindrance in the penetration of organic bioactive compounds into the solution, ultimately resulting in a decrease in the yield of the extraction process.

3.4.3. Effect of Solvent Concentration

The solubility of organic bioactive compounds can be enhanced by varying the concentration of the solvent [39]. Olive leaves were extracted using different ethanol concentrations at no more than 50 °C for 60 min under ultrasonic conditions. The results, shown in Figure 4, indicate that the samples containing 60% ethanol had the highest TPC, TFC, DPPH, ABTS, and FRAP values. This is because the solubility of polyphenol compounds increases with increasing ethanol concentrations. This may explain why the presence of water in ethanol improves the swelling of the plant material, while ethanol disrupts the binding between the solute and the plant. Other studies [40,41] have found that the recovery of organic bioactive compounds increased and peaked at an ethanol concentration of around 70% before slightly decreasing. Studies have shown that increasing the ethanol concentration can enhance the yield of phenolic compounds up to an average concentration of 40% ethanol [42]; however, Caldas et al. [43] observed that the highest phenolic compound content was achieved at an average concentration of 60% ethanol, possibly due to the different polarities of organic bioactive compounds in grape skin. For *Triticum aestivum*, the maximum yield of phenolic compounds was obtained using an ethanol concentration of 56% [44]. Similarly, studies on various plants such as *Jaboticaba* bark, blueberry, and apple pulp have found that the concentration of ethanol required to exstract the maximum amount of phenolic compounds ranges from 40–80% (V/V) [35–46].

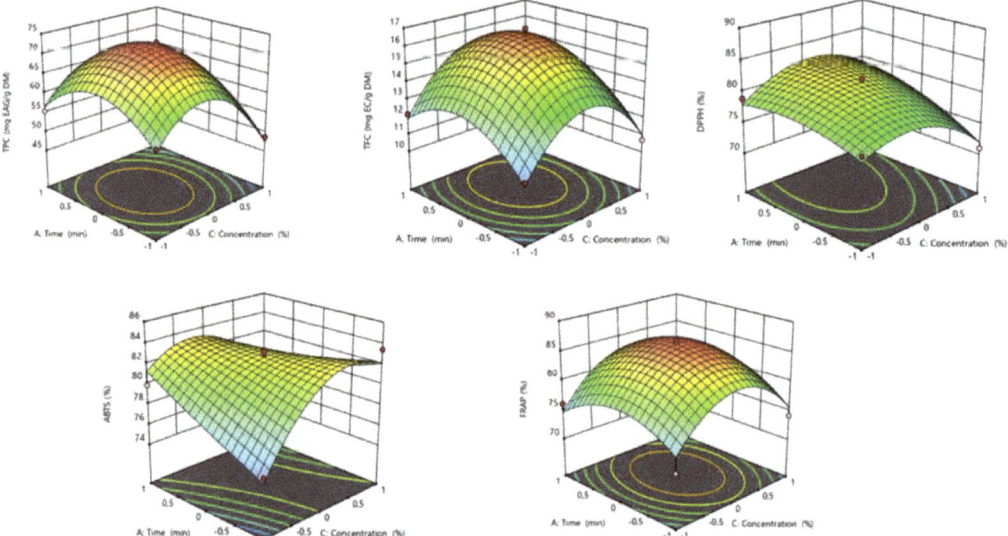

Figure 4. Response surface plot showing the variation of responses with ethanol concentration.

3.5. Determination and Validation of Optimal Conditions

This study aimed to determine the best experimental conditions for extracting phenolic compounds that benefit human health, such as antioxidant activity. This involves determining the ideal sonication time, solid/liquid ratio, and ethanol concentration for maximum extraction yield. The software Design Expert Version 13 was used to find the composite optimum for achieving the highest bioactive compounds extraction yield and the most significant antioxidant activity.

The optimization process involved assigning the response's desirability values between 0 and 1. Figure 5 displays a desirability value of 0.8735 and the predicted optimal values. Experiments were conducted under optimal triplicate conditions to compare the experimental and predicted values of the responses. The mean values are presented in Table 4. The optimal conditions for the experiments were 53.52 min, a solid/liquid ratio of 9.83 mL/g, and an ethanol concentration of 59.7% (V/V). The experimental values for TPC and TFC bioactive compounds were 74.45 ± 1.22 mg EAG/g DM and 17.08 ± 1.85 mg EC/g DM, respectively, while the antioxidant activity DPPH, ABTS, and FRAP were 83.45 ± 0.89%, 82.85 ± 1.52%, and 87.01 ± 2.35%, respectively. These experimental results were found to be similar to the predicted model for TPC (72.40 mg EAG/g DM), TFC (16.42 mg EC/g DM), DPPH (82.58%), ABTS (83.06%), and FRAP (85.79%). Therefore, there is a synergy between the results found and the Box–Behnken design.

Table 4. Optimal conditions and predicted and actual response values of olive leaf extract.

The Optimal Conditions			The Answers							
X1 (min)	X2 (mL/g)	X3 (%)	TPC (mg GAE/g DM)		TFC (mg EC/g DM)		DPPH (%)		ABTS (%)		FRAP (%)	
			Reel	Predict	Reel	Predict	Reel	Predict	Reel	Predict	Reel	Predict
53.5	9.83	59.7	Maximum		Maximum		Maximum		Maximum		Maximum	
Objective Optimized values			74.45 ± 1.22	72.40	17.08 ± 1.85	16.42	83.45 ± 0.89	82.58	82.85 ± 1.52	83.06	87.01 ± 2.35	85.79

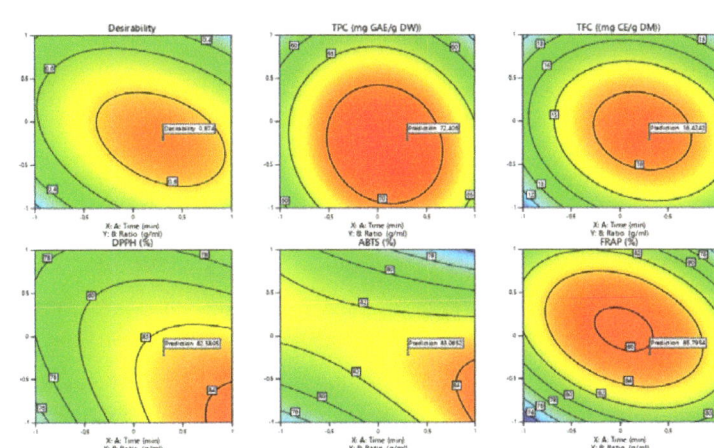

Figure 5. The predicted optimal values and desirability of olive leaf optimization.

3.6. HPLC-MS Analysis

Figure 6 displays the phenolic compounds identified in olive leaf extracts through ultrasound in the first and second extractions. The peaks on the graph correspond to these compounds. The results indicate that the first extraction in optimal conditions (Figure 6a) resulted in a high release of phenolic compounds. In contrast, the second extraction (Figure 6b) had fewer bioactive compounds. This suggests that the ultrasonic extraction method is effective. Eight phenolic compounds were identified: hydroxytyrosol, catechin, caffeic acid, vanillin, naringin, oleuropein, quercetin, and kaempferol. As shown in Table 5, oleuropein was the most abundant compound, with a concentration of 114.10 mg/g DM, followed by hydroxytyrosol, caffeic acid, and kaempferol. The optimized ultrasound-assisted extraction method was likely responsible for the high amount of phenolic compounds in the first extraction.

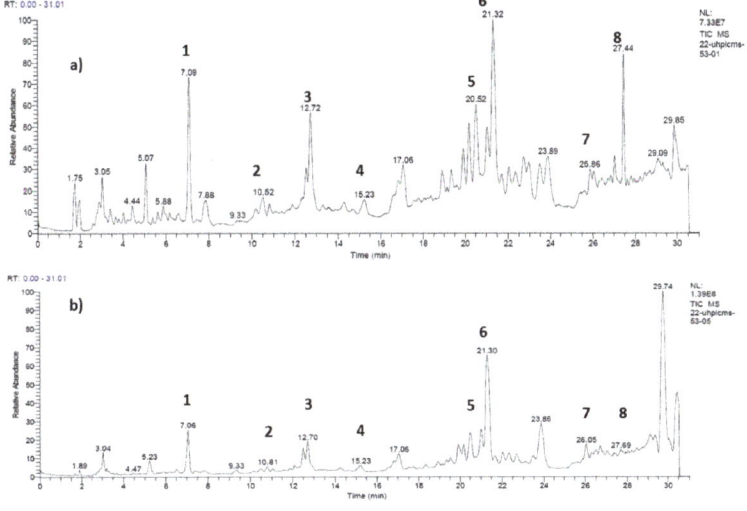

Figure 6. HPLC chromatograms of polyphenols of olive leaf extracts from the First UEA under the optimal conditions (**a**) and the Second UEA under the same optimal conditions (**b**).

Table 5. Concentration of phenolic compounds identified in olive leaf extracts under the optimal conditions of the first and second ultrasound extraction (mg/g DM).

N° PIC	T_R (min)	Phenolic Compounds	Chemical Formulas	Concentration of the First Extraction (mg/g)	Concentration of Second Extraction (mg/g)
1	7.09	Hydroxytyrosol	$C_8H_{10}O_3$	45.40 ± 1.2	10.02 ± 2.12
2	10.52	Catechin	$C_{15}H_{14}O_6$	12.90 ± 1.35	2.21 ± 1.36
3	12.72	Caffeic acid	$C_9H_8O_4$	79.50 ± 1.25	15.32 ± 3.36
4	15.23	Vanillin	$C_8H_8O_3$	12.70 ± 1.36	0.98 ± 2.36
5	20.52	Naringin	$C_9H_8O_3$	45.40 ± 2.45	32.23 ± 1.25
6	21.30	Oleuropein	$C_{25}H_{32}O_{13}$	114.10 ± 3.42	40.23 ± 2.78
7	26.48	Quercetin	$C_{15}H_{10}O_7$	23.00 ± 2.38	12.21 ± 1.45
8	27.44	Kaempferol	$C_{15}H_{10}O_6$	29.00 ± 1.96	9.36 ± 1.69

3.7. Scanning Electron Microscopy

After being dried and ground, the leaf powder was examined under a scanning electron microscope. There was a visible difference between the untreated sample and the one treated with EWM and UAE (as shown in Figure 7). The untreated sample was densely compacted, while the treated sample showed structural changes. UAE caused more damage and formation of cracks than maceration, possibly due to the cavitation effects of the ultrasound [47]. During extraction, high ultrasound intensities can enhance solvent penetration and destroy cell membranes [48], releasing more bioactive compounds from the sample matrix.

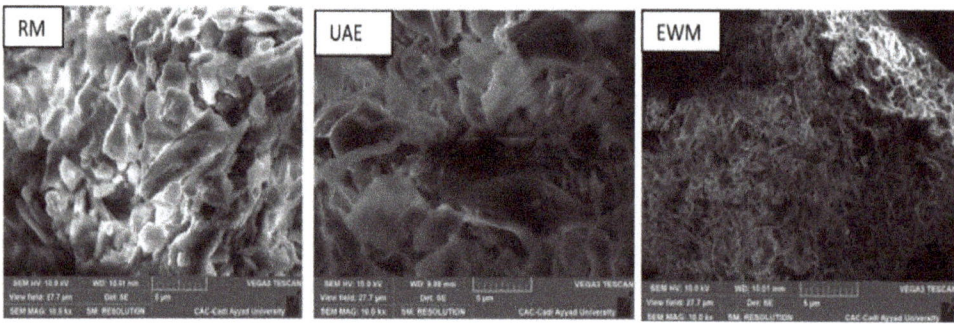

Figure 7. SEM images of olive leaf powder before extraction (RM), leaf powder after ultrasound-assisted extraction (UAE) under optimal conditions, and leaf powder treated by ethanol water maceration (EWM).

4. Conclusions

The Box–Behnken design (BBD) method, along with the surface response design approach (RSM), was used to study the impact of ultrasonic-assisted extraction (UAE) process parameters on the content of polyphenolic compounds in *Moroccan picholine* olive leaves. The results showed that this eco-friendly technique is beneficial in optimizing the conditions for extracting phenolic compounds (TPC, TFC). To achieve a high yield, it is recommended to use an extraction time of 53.5 min, a solvent/solid ratio of 9.95 mL/g, and an ethanol concentration of 59.7%. The content of TPC and TFC are 74.45 ± 1.22 mg EAG/g DM and 17.08 ± 1.85 mg EC/g DM, respectively, while the antioxidant activity of DPPH, ABTS, and FRAP are 83.45 ± 0.89%, 82.85 ± 1.52%, and 87.01 ± 2.35% respectively. The presence of certain phenolic bioactive compounds in high concentrations, specifically oleuropein and hydroxytyrosol, was confirmed through analysis by HPLC-MS.

To better understand the extraction and optimization phenomena, it would be beneficial to study additional parameters such as sonication temperature, frequency, and solvent nature.

Author Contributions: E.M.E.A.: doctoral student, practical work, and main leader of the operative manipulations; A.M.: doctoral assistant of operations and preparation of raw material; N.E. (FSSM/UCA), A.J. (FSSM/UCA) and M.M. (Emeritus FSSM/UCA): professors who initiated the follow-up and planning of the thesis work; M.O. and N.H.: assistants doctors of analysis; K.B., doctor, pharmacist, and CEO of the company, Iris Cosmétologie (IRCOS) Laboratoires, our industrial partner: availability of raw materials, logistics and production machines. All authors have read and agreed to the published version of the manuscript.

Funding: This research was funded by the National Center for Scientific and Technical Research (CNRST) and the Ministry of Higher Education, Scientific Research and Vocational Training of Morocco, "PPR-BR2BINOV-Mahrouz-FS-UCA-Marrakesh".

Data Availability Statement: Not applicable.

Acknowledgments: The authors thank the City of Innovation of Marrakech and the company, Iris Cosmétologie (IRCOS) Laboratoires.

Conflicts of Interest: The authors declare no conflict of interest.

References

1. Prakash Maran, J.; Manikandan, S.; Thirugnanasambandham, K.; Vigna Nivetha, C.; Dinesh, R. Box–Behnken Design Based Statistical Modeling for Ultrasound-Assisted Extraction of Corn Silk Polysaccharide. *Carbohydr. Polym.* **2013**, *92*, 604–611. [CrossRef] [PubMed]
2. Benavente-García, O.; Castillo, J.; Lorente, J.; Ortuño, A.; Del Rio, J.A. Antioxidant Activity of Phenolics Extracted from Olea Europaea L. Leaves. *Food Chem.* **2000**, *68*, 457–462. [CrossRef]
3. Ghomari, O.; Sounni, F.; Massaoudi, Y.; Ghanam, J.; Drissi Kaitouni, L.B.; Merzouki, M.; Benlemlih, M. Phenolic Profile (HPLC-UV) of Olive Leaves According to Extraction Procedure and Assessment of Antibacterial Activity. *Biotechnol. Rep.* **2019**, *23*, e00347. [CrossRef] [PubMed]
4. Jemai, H.; Bouaziz, M.; Fki, I.; El Feki, A.; Sayadi, S. Hypolipidimic and Antioxidant Activities of Oleuropein and Its Hydrolysis Derivative-Rich Extracts from Chemlali Olive Leaves. *Chem.-Biol. Interact.* **2008**, *176*, 88–98. [CrossRef]
5. Niaounakis, M.; Halvadakis, C.P. *Waste Management Series*; Elsevier: Amsterdam, The Netherlands, 2006; Volume 5, ISBN 978-0-08-044851-0.
6. Fares, R.; Bazzi, S.; Baydoun, S.E.; Abdel-Massih, R.M. The Antioxidant and Anti-Proliferative Activity of the Lebanese Olea Europaea Extract. *Plant Foods Hum. Nutr.* **2011**, *66*, 58–63. [CrossRef]
7. Vian, M.A.; Fernandez, X.; Visinoni, F.; Chemat, F. Microwave Hydrodiffusion and Gravity, a New Technique for Extraction of Essential Oils. *J. Chromatogr. A* **2008**, *1190*, 14–17. [CrossRef]
8. Sirichan, T.; Kijpatanasilp, I.; Asadatorn, N.; Assatarakul, K. Optimization of Ultrasound Extraction of Functional Compound from Makiang Seed by Response Surface Methodology and Antimicrobial Activity of Optimized Extract with Its Application in Orange Juice. *Ultrason. Sonochem.* **2022**, *83*, 105916. [CrossRef]
9. Wen, C.; Zhang, J.; Zhang, H.; Dzah, C.S.; Zandile, M.; Duan, Y.; Ma, H.; Luo, X. Advances in Ultrasound Assisted Extraction of Bioactive Compounds from Cash Crops—A Review. *Ultrason. Sonochem.* **2018**, *48*, 538–549. [CrossRef]
10. Saifullah, M.; McCullum, R.; McCluskey, A.; Vuong, Q. Comparison of Conventional Extraction Technique with Ultrasound Assisted Extraction on Recovery of Phenolic Compounds from Lemon Scented Tea Tree (*Leptospermum Petersonii*) Leaves. *Heliyon* **2020**, *6*, e03666. [CrossRef]
11. Al-Dhabi, N.A.; Ponmurugan, K.; Maran Jeganathan, P. Development and Validation of Ultrasound-Assisted Solid-Liquid Extraction of Phenolic Compounds from Waste Spent Coffee Grounds. *Ultrason. Sonochem.* **2017**, *34*, 206–213. [CrossRef]
12. Baş, D.; Boyacı, İ.H. Modeling and Optimization I: Usability of Response Surface Methodology. *J. Food Eng.* **2007**, *78*, 836–845. [CrossRef]
13. Bezerra, M.A.; Santelli, R.E.; Oliveira, E.P.; Villar, L.S.; Escaleira, L.A. Response Surface Methodology (RSM) as a Tool for Optimization in Analytical Chemistry. *Talanta* **2008**, *76*, 965–977. [CrossRef] [PubMed]
14. Hou, X.; Tang, S.; Guo, X.; Wang, L.; Liu, X.; Lu, X.; Guo, Y. Preparation and Application of Guanidyl-Functionalized Graphene Oxide-Grafted Silica for Efficient Extraction of Acidic Herbicides by Box-Behnken Design. *J. Chromatogr. A* **2018**, *1571*, 65–75. [CrossRef] [PubMed]
15. Mourjane, A.; Hanine, H.; El Adnany, E.M.; Ouhammou, M.; Hidar, N.; Nabil, B.; Boumendjel, A.; Bitar, K.; Mahrouz, M. Energetic Bio-Activation of Some Organic Molecules and Their Antioxidant Activity in the Pulp of the Moroccan Argan Tree «*Argania spinosa* L.». *Molecules* **2022**, *27*, 3329. [CrossRef]
16. Dai, J.; Mumper, R.J. Plant Phenolics: Extraction, Analysis and Their Antioxidant and Anticancer Properties. *Molecules* **2010**, *15*, 7313–7352. [CrossRef] [PubMed]
17. Chemat, F.; Vian, M.A.; Cravotto, G. Green Extraction of Natural Products: Concept and Principles. *IJMS* **2012**, *13*, 8615–8627. [CrossRef]
18. Clary, J.J. *The Toxicology of Methanol*; John Wiley & Sons: Hoboken, NJ, USA, 2013; ISBN 0-470-31759-0.

19. Ramić, M.; Vidović, S.; Zeković, Z.; Vladić, J.; Cvejin, A.; Pavlić, B. Modeling and Optimization of Ultrasound-Assisted Extraction of Polyphenolic Compounds from Aronia Melanocarpa by-Products from Filter-Tea Factory. *Ultrason. Sonochem.* **2015**, *23*, 360–368. [CrossRef]
20. Slinkard, K.; Singleton, V.L. Total Phenol Analysis: Automation and Comparison with Manual Methods. *Am. J. Enol. Vitic.* **1977**, *28*, 49–55. [CrossRef]
21. Djeridane, A.; Yousfi, M.; Nadjemi, B.; Maamri, S.; Djireb, F.; Stocker, P. Phenolic Extracts from Various Algerian Plants as Strong Inhibitors of Porcine Liver Carboxylesterase. *J. Enzym. Inhib. Med. Chem.* **2006**, *21*, 719–726. [CrossRef]
22. Abdel-Hameed, E.-S.S.; Nagaty, M.A.; Salman, M.S.; Bazaid, S.A. Phytochemicals, Nutritionals and Antioxidant Properties of Two Prickly Pear Cactus Cultivars (Opuntia Ficus Indica Mill.) Growing in Taif, KSA. *Food Chem.* **2014**, *160*, 31–38. [CrossRef]
23. Aadesariya, M.K.; Ram, V.R.; Dave, P.N. Evaluation of Antioxidant Activities by Use of Various Extracts from Abutilon Pannosum and Grewia Tenax Leaves in the Kachchh Region. *MOJ Food Process. Technol.* **2017**, *17*, 359.
24. Oyaizu, M. Studies on Products of Browning Reaction Antioxidative Activities of Products of Browning Reaction Prepared from Glucosamine. *Jpn. J. Nutr. Diet.* **1986**, *44*, 307–315. [CrossRef]
25. Soltani, Y.; Ali-Bouzidi, M.; Toumi, F.; Benyamina, A. Activités Antioxydantes Des Extraits de Trois Organes de Juniperus Phoenicea L. de l'Ouest Algérien. *Phytothérapie* **2017**, *16*, 142–148. [CrossRef]
26. Benzie, I.F.; Strain, J.J. The Ferric Reducing Ability of Plasma (FRAP) as a Measure of "Antioxidant Power": The FRAP Assay. *Anal. Biochem.* **1996**, *239*, 70–76. [CrossRef]
27. Puigventós, L.; Navarro, M.; Alechaga, É.; Núñez, O.; Saurina, J.; Hernández-Cassou, S.; Puignou, L. Determination of Polyphenolic Profiles by Liquid Chromatography-Electrospray-Tandem Mass Spectrometry for the Authentication of Fruit Extracts. *Anal. Bioanal. Chem.* **2015**, *407*, 597–608. [CrossRef]
28. Harborne, A. *Phytochemical Methods: A Guide to Modern Techniques of Plant Analysis*; Springer Science & Business Media: Berlin/Heidelberg, Germany, 1998; ISBN 0-412-57270-2.
29. Atkinson, A.C.; Donev, A.N. *Optimum Experimental Designs*; Clarendon Press: Oxford, UK, 1992; Volume 5.
30. Chakraborty, S.; Uppaluri, R.; Das, C. Optimization of Ultrasound-Assisted Extraction (UAE) Process for the Recovery of Bioactive Compounds from Bitter Gourd Using Response Surface Methodology (RSM). *Food Bioprod. Process.* **2020**, *120*, 114–122. [CrossRef]
31. Tiwari, B.K.; O'Donnell, C.P.; Cullen, P.J. Effect of Non Thermal Processing Technologies on the Anthocyanin Content of Fruit Juices. *Trends Food Sci. Technol.* **2009**, *20*, 137–145. [CrossRef]
32. Jovanović, A.A.; Đorđević, V.B.; Zdunić, G.M.; Pljevljakušić, D.S.; Šavikin, K.P.; Gođevac, D.M.; Bugarski, B.M. Optimization of the Extraction Process of Polyphenols from Thymus Serpyllum L. Herb Using Maceration, Heat- and Ultrasound-Assisted Techniques. *Sep. Purif. Technol.* **2017**, *179*, 369–380. [CrossRef]
33. Li, F.; Raza, A.; Wang, Y.-W.; Xu, X.-Q.; Chen, G.-H. Optimization of Surfactant-Mediated, Ultrasonic-Assisted Extraction of Antioxidant Polyphenols from Rattan Tea (*Ampelopsis grossedentata*) Using Response Surface Methodology. *Phcog. Mag.* **2017**, *13*, 446. [CrossRef]
34. Madrona, G.S.; Terra, N.M.; Coutinho Filho, U.; de Santana Magalhães, F.; Cardoso, V.L.; Reis, M.H.M. Purification of Phenolic Compounds from Genipap (*Genipa americana* L.) Extract by the Ultrasound Assisted Ultrafiltration Process. *Acta Scientiarum. Technol.* **2019**, *41*, e35571. [CrossRef]
35. Rocha, J.D.C.G.; Procopio, F.R.; Mendonca, A.C.; Vieira, L.M.; Perrone, I.T.; Barros, F.A.R.D.; Stringheta, P.C. Optimization of Ultrasound-Assisted Extraction of Phenolic Compounds from Jussara (*Euterpe edulis* M.) and Blueberry (*Vaccinium myrtillus*) Fruits. *Food Sci. Technol.* **2017**, *38*, 45–53. [CrossRef]
36. Saci, F.; Benchikh, Y.; Louaileche, H.; Bey, M.B. Optimization of Ultrasound-Assisted Extraction of Phenolic Compounds and Antioxidant Activity from Carob Pulp (*Ceratonia siliqua* l.) by Using Response Surface Methodology. *Ann. Univ. Dunarea Jos Galati. Fascicle VI-Food Technol.* **2018**, *42*, 26–39.
37. Xiao, W.; Han, L.; Shi, B. Microwave-Assisted Extraction of Flavonoids from Radix Astragali. *Sep. Purif. Technol.* **2008**, *62*, 614–618. [CrossRef]
38. Zakaria, F.; Tan, J.-K.; Mohd Faudzi, S.M.; Abdul Rahman, M.B.; Ashari, S.E. Ultrasound-Assisted Extraction Conditions Optimisation Using Response Surface Methodology from Mitragyna Speciosa (Korth.) Havil Leaves. *Ultrason. Sonochem.* **2021**, *81*, 105851. [CrossRef] [PubMed]
39. Tubtimdee, C.; Shotipruk, A. Extraction of Phenolics from Terminalia Chebula Retz with Water–Ethanol and Water–Propylene Glycol and Sugaring-out Concentration of Extracts. *Sep. Purif. Technol.* **2011**, *77*, 339–346. [CrossRef]
40. He, B.; Zhang, L.-L.; Yue, X.-Y.; Liang, J.; Jiang, J.; Gao, X.-L.; Yue, P.-X. Optimization of Ultrasound-Assisted Extraction of Phenolic Compounds and Anthocyanins from Blueberry (*Vaccinium ashei*) Wine Pomace. *Food Chem.* **2016**, *204*, 70–76. [CrossRef]
41. Noroozi, F.; Bimakr, M.; Ganjloo, A.; Aminzare, M. A Short Time Bioactive Compounds Extraction from Cucurbita Pepo Seed Using Continuous Ultrasound-assisted Extraction. *Food Meas.* **2021**, *15*, 2135–2145. [CrossRef]
42. Kumar, K.; Srivastav, S.; Sharanagat, V.S. Ultrasound Assisted Extraction (UAE) of Bioactive Compounds from Fruit and Vegetable Processing by-Products: A Review. *Ultrason. Sonochem.* **2021**, *70*, 105325. [CrossRef]
43. Caldas, T.W.; Mazza, K.E.L.; Teles, A.S.C.; Mattos, G.N.; Brígida, A.I.S.; Conte-Junior, C.A.; Borguini, R.G.; Godoy, R.L.O.; Cabral, L.M.C.; Tonon, R.V. Phenolic Compounds Recovery from Grape Skin Using Conventional and Non-Conventional Extraction Methods. *Ind. Crops Prod.* **2018**, *111*, 86–91. [CrossRef]

44. Savic, I.M.; Savic Gajic, I.M. Optimization of Ultrasound-Assisted Extraction of Polyphenols from Wheatgrass (*Triticum aestivum* L.). *J. Food Sci. Technol.* **2020**, *57*, 2809–2818. [CrossRef]
45. Rodrigues, S.; Fernandes, F.A.N.; de Brito, E.S.; Sousa, A.D.; Narain, N. Ultrasound Extraction of Phenolics and Anthocyanins from Jabuticaba Peel. *Ind. Crops Prod.* **2015**, *69*, 400–407. [CrossRef]
46. Zvicevičius, G.; Liaudanskas, M.; Viškelis, P. Phenolic Compound Quantification and Antioxidant Activity Determination in Ethanol Extracts of Apple Pulp and Peels. 2014. Available online: https://lsmu.lt/cris/handle/20.500.12512/15459 (accessed on 14 April 2023).
47. Kaderides, K.; Papaoikonomou, L.; Serafim, M.; Goula, A.M. Microwave-Assisted Extraction of Phenolics from Pomegranate Peels: Optimization, Kinetics, and Comparison with Ultrasounds Extraction. *Chem. Eng. Process.-Process Intensif.* **2019**, *137*, 1–11. [CrossRef]
48. Li, H.; Pordesimo, L.; Weiss, J. High Intensity Ultrasound-Assisted Extraction of Oil from Soybeans. *Food Res. Int.* **2004**, *37*, 731–738. [CrossRef]

Disclaimer/Publisher's Note: The statements, opinions and data contained in all publications are solely those of the individual author(s) and contributor(s) and not of MDPI and/or the editor(s). MDPI and/or the editor(s) disclaim responsibility for any injury to people or property resulting from any ideas, methods, instructions or products referred to in the content.

Article

Fractions of Methanol Extracts from the Resurrection Plant *Haberlea rhodopensis* Have Anti-Breast Cancer Effects in Model Cell Systems

Diana Zasheva [1,*], Petko Mladenov [2], Krasimir Rusanov [2], Svetlana Simova [3], Silvina Zapryanova [1], Lyudmila Simova-Stoilova [4], Daniela Moyankova [2] and Dimitar Djilianov [2]

1. Institute of Biology and Immunology of Reproduction, Bulgarian Academy of Sciences, Tsarigradsko Shosse, 73, 1113 Sofia, Bulgaria; silvina_z@abv.bg
2. Agrobioinstitute, Agricultural Academy, 1164 Sofia, Bulgaria; mladenovpetko@yahoo.com (P.M.); krusanov@abv.bg (K.R.); dmoyankova@abi.bg (D.M.); d_djilianov@abi.bg (D.D.)
3. Institute of Organic Chemistry with Centre of Phytochemistry (IOCCP), Bulgarian Academy of Sciences, 1113 Sofia, Bulgaria; svetlana.simova@orgchm.bas.bg
4. Institute of Plant Physiology and Genetics, Bulgarian Academy of Science, 1113 Sofia, Bulgaria; lpsimova@yahoo.co.uk
* Correspondence: zasheva.diana@yahoo.com

Abstract: Breast cancer is among the most problematic diseases and a leading cause of death in women. The methods of therapy widely used, so far, are often with many side effects, seriously hampering patients' quality of life. To overcome these constraints, new cancer treatment alternatives are constantly tested, including bioactive compounds of plant origin. Our aim was to study the effects of *Haberlea rhodopensis* methanol extract fractions on cell viability and proliferation of two model breast cancer cell lines with different characteristics. In addition to the strong reduction in cell viability, two of the fractions showed significant influence on the proliferation rate of the hormone receptor expressing MCF7 and the triple negative MDA-MB231 breast cancer cell lines. No significant effects on the benign MCF10A cell line were observed. We applied a large scale non-targeted approach to purify and identify highly abundant compounds from the active fractions of *H. rhodopensis* extracts. By the combined NMR/MS approach, myconoside was identified in the fractions and hispidulin 8-C-(6-O-acetyl-2″-O-syringoyl-β-glucopyranoside) was found in one of them. We further performed molecular docking analysis of possible myconoside interactions with several proteins, important for breast cancer proliferation. High probability of binding was established for GLUT1 transporter, estrogen receptor and MYST acetyltransferase. Our results are a good background for future studies on the use of myconoside for targeted breast cancer therapy.

Keywords: breast cancer; *Haberlea rhodopensis*; myconoside; hispidulin 8-C-(6-O-acetyl-2″-O-syringoyl-β-glucopyranoside); GLUT1 transporter; estrogen receptor and MYST acetyltransferase

Citation: Zasheva, D.; Mladenov, P.; Rusanov, K.; Simova, S.; Zapryanova, S.; Simova-Stoilova, L.; Moyankova, D.; Djilianov, D. Fractions of Methanol Extracts from the Resurrection Plant *Haberlea rhodopensis* Have Anti-Breast Cancer Effects in Model Cell Systems. *Separations* **2023**, *10*, 388. https://doi.org/10.3390/separations10070388

Academic Editor: Faiyaz Shakeel

Received: 6 June 2023
Revised: 27 June 2023
Accepted: 28 June 2023
Published: 1 July 2023

Copyright: © 2023 by the authors. Licensee MDPI, Basel, Switzerland. This article is an open access article distributed under the terms and conditions of the Creative Commons Attribution (CC BY) license (https:// creativecommons.org/licenses/by/ 4.0/).

1. Introduction

One of the most problematic diseases related to women's health is breast cancer. Cases of breast cancer diagnosed in 2008 were 1.38 million [1] and their number increased to 2.3 million in 2020 [2], thus reaching 12% of all cancer cases [3]. At the same time, breast cancer is the second leading cause of death in women [4]. The first step in finding new anticancer substances is to test them on model cell lines that have features common to different types of cancer. Cell cultures remain indispensable tools in cancer research, despite some limitations due to phenotypic drifts, some heterogeneity and existence of clonal variants. The cellular characteristics of breast cancer and the changes in their cell signal pathways complicate the therapeutic methods used so far. Invasive cancer types, named basal, are characterized with low expression of HER (Human Epidermal Growth Factor Receptor) and a loss of estrogen and progesterone receptors [5]. They are also known as triple negative and are unresponsive of hormonal replacement therapy. A good model system to study new anticancer agents suitable for this very aggressive type of cancer is the cell line MDA-MB231 [6]. The expression of HER2 receptor and two hormonal receptors (estrogen and progesterone ones) characterizes luminal B types of breast cancer, known also as triple positive. A widely used model system of the hormone receptor expressing cancer is the cell line MCF7 [4]. This cell line is also used to study epigenetic regulation of cancer growth, mediated by higher expression and activity of MYST acetyltransferases in estrogen dependent breast tumors [7]. Several breast cancer cell lines, including MDA-MB231 and MCF7, are characterized with high rates of glucose uptake and high expression of glucose transporters of the GLUT family [6], which is not typical for noncancerous epithelial cell lines such as MCF-10A.

Conventionally used therapeutic methods are invasive and with many side effects. Often, the standardly used chemotherapeutic drugs lose effectiveness because of multidrug resistance developed by cancer cells. Side effects of chemotherapy and radiotherapy seriously hamper patients' quality of life. To overcome these constraints, new cancer treatment alternatives are constantly tested, including bioactive compounds of plant origin [8–10]. Polyphenol substances like e.g., coumarins [11], flavones like genistein [12], phenols like thymol [13], monoterpenes like thymoquinone [14] have been found to reduce the viability of breast cancer cell lines of various origins by mechanisms related to switching on apoptotic pathways, by blocking cell proliferation or different kinase pathways. The development of analytics with high resolution complementary instruments allows identification and determination of the molecular structure of many new active compounds from various plant species. In this respect, the complementary data obtained by Nuclear Magnetic Resonance (NMR) and Mass Spectrometry (MS) could provide reliable information for compound discovery in natural products research [15,16]. In addition, identification of bioactive compounds from various plant species is a good perspective for drug discovery, with the valuable contribution of Artificial Intelligence (AI) and synthetic chemistry, for therapy or even prevention of cancer. In the search for new therapeutic agents, there has been recent attention on the resurrection plant species. They are a group of higher plants that are able to withstand a drastic decrease in the water content of their vegetative tissues and after long periods of dryness, they are able to recover fast and fully when water is available again [17]. These plant species belong to various botanical families and live under differing environments but share high desiccation tolerance as a common characteristic. This feature makes resurrection plants a suitable model for intensive studies for stress tolerance at molecular, physiological, biochemical and metabolomics levels. The antioxidative component of resurrection plants' desiccation tolerance is well recognized [18]. Additional attention is paid to the specific secondary metabolites, constitutive or accumulated during stress, with potential application as food additives, cosmetic agents or medicinal components.

The Balkan endemite *Haberlea rhodopensis* is among the only few resurrection plants growing in Europe and as all other species on the continent, belongs to the Gesneriaceae botanical family [19]. It is among the most studied desiccation tolerant model systems, at whole plant level or as detached leaves assays [20–25].

In addition, following long established strategy [26], knowledge on the specific metabolome of *H. rhodopensis* was gradually generated and the potential application of extracts or isolated compounds in various areas has been studied, including human diseases [27–31].

The present study is the first attempt to follow the behavior of two breast cancer cell lines with different characteristics in comparison to a normal breast epithelial cell line after treatment with *H. rhodopenis* extracts and their fractions. The latter are resulting from a non-targeted approach to purify and identify highly abundant compounds involved in the significant reduction in cell viability and proliferation of the cancer cell lines. Molecular docking analysis has been performed on one of the identified compounds—myconoside to propose a model for its interaction with several cancer proteins in an attempt to explain the potential mechanisms of its penetration in cells and the reduction in their viability and proliferation. The results are discussed as a background for further studies and potential applications.

2. Materials and Methods

2.1. Chemicals and Reagents

Acetonitril and methanol of HPLC grade from Macron Fine Chemicals™ (Avantor, Gliwice, Poland) and Acetonitril of LC/MS grade from Sigma-Aldrich (St. Louis, MO, USA) were used. Dimethyl sulfoxide (DMSO) and formic acid were sourced in analytical grade from Sigma-Aldrich (St. Louis, MO, USA). Sephadex LH-20 was purchased from Cytiva (Marlborough, MA, USA). Antibiotic/antimycotic solution was purchased from GE Healthcare (Boston, MA, USA). CD3OD was purchased from Euriso-top (Saint-Aubin, France). All reagents for cell cultures treatments and cell culture assay tests were purchased from the Sigma-Aldrich (St. Louis, MO, USA).

2.2. Plant Material and Leaf Extract Preparation and Fractionation by Size Exclusion Chromatography

H. rhodopensis plants were propagated in vitro and adapted in pots under controlled greenhouse conditions at 22–24 °C, a 16-h photoperiod, 60% relative humidity and a photon flux density of 36 $\mu mol m^{-2} s^{-1}$ [32]. Fully developed leaves from well-hydrated pot plants were detached and air-dried for methanol extract preparation. Homogenised leaves (6 g) were macerated in 60 mL methanol and extracted for 30 min at 70 °C on water bath. After centrifugation at $10,000 \times g$ for 10 min, the supernatants were collected, dried at 40 °C by SpeedVac Labconco (Kansas City, MI, USA) and stored at -20 °C for further use.

1.1276 g of crude extract were dissolved in 14 mL methanol and subjected to size exclusion chromatography (SEC) on a Cytiva XK 50/100 column filled with Sephadex LH-20 with the help of an AKTA Pure FPLC system (Cytiva) using a constant flow of 3 mL/min of methanol. Monitoring of the elution was carried out at 220 nm. SEC fractions were collected in 50 mL tubes based on the observed elution of UV absorbing compounds.

2.3. Cell Cultures Treatments and Cell Culture Assay Tests

2.3.1. Cell Lines

In this study adherent breast cancer cell line MCF7 (ATCC cell culture collection N NTB-22TM) Manassas, Virginia, USA and MDA-MB231 (ATCC cell culture collection N HTB-26™) and normal adherent breast epithelial cell line MCF 10A (ATCC cell culture collection N CRL-10317™) Manassas, Virginia, USA) were used. The normal cell line was grown on DMEM/F12 medium and MDA-MB231 cancer cell line was grown on DMEM medium with high glucose (4.5 g/L) supplemented with 5% FBS and 10% FBS, respectively, and antibiotic/antimycotic solution Amino acid mix solution was added to the cancer cell line MDA-MB231 medium. The normal cell line needed insulin in concentration 10 µg/mL, endothelial growth factor in concentration 20 ng/mL and hydrocortisone in concentration 500 ng/mL. The cancer cell line MCF7 was incubated in DMEM with low glucose (1 g/L) supplemented with 10% FBS and antibiotic/antimycotic solution and insulin in concentration 0.01 mg/mL. Cells were incubated in 5% CO_2 incubator at 37 °C. The cells were cultivated to 80–90% confluence and were trypsinized with trypsin/EDTA solution. The cells were centrifuged, suspended in in FBS with 5–10% DMSO and stored in freezer at -150 °C.

2.3.2. MTT Cell Viability Assay

The studied cell lines were grown in 25 cm^2 well plates to confluence 80–85%. 10^4 cells were seeded per 96 well plate. The cells were grown 24 h and then treated with *H. rhodopensis* total extract (fractions). The viability was determined 48 h after the treatment with total extract and fractions at 48th, 72nd hour. MTT test followed standard procedure [33]—20 µL of MTT stock solution (5 mg/mL) were added to the well and the cells were incubated in CO₂ incubator for 4 h. The formazan crystals were solubilized in 150 µL dimethyl sulphoxide. The color intensity was measured at wave length 600 nm (ELISA Reader Fluostar Optima/BMG Tech) ThermoFisher Scientific Corporation, Waltham, MA, USA. The untreated samples were used as control with 100% viability. The percent of viable cells in experimental conditions were scored as a percent of control sample.

% of viable cells = E treated sample/E control sample × 100)

Each condition in experiment was performed in triplicate. Three biological experiments were made for each of the cell lines. The standard error bars in percent are presented in the graphs. The total extract was applied in concentrations 10, 100, 200 and 300 µg/mL, respectively, and the fractions were applied in concentration 300 µg/mL.

2.3.3. Proliferation Assay

The trypan blue-excluding proliferation assay test was made following the next procedure [33]. The cells were grown in 24-well plate to 4.5×10^4 cell for MCF10-A cell line, to 6×10^4 cells for MCF7 cell line and to 4×10^4 cells for MDA-MB231 cell line. For treatment, the medium was changed with fresh medium containing *H. rhodopensis* fractions numbers 14 and 18 at concentration 300 µg/mL. Total number of cells was counted for 96 h at 24 h interval. The cells were washed with PBS and then they were suspended in PBS, 10–20 µL of their suspension was trypan blue stained. Three biological replicates were made. The living cells presented in the graphs are averages of total number of cells in well minus averages of number of dead cells in well, stained by trypan blue dye.

2.3.4. Statistical Methods

The results are presented with standard error bars [34]. The absolute values of MTT assay of breast cancer cell lines data for total extract treatment concentration 300 µg/ml and fractions with effect on cancer cell lines are processed in ANOVA for multiple comparison plot and the data are presented extrapolated in absolute values to log2 scale. For each of the groups was applied one-way ANOVA analysis of fractions treatment versus control at each time point of the studied normal and two cancer cell lines in proliferation assay to estimate variations within the group [35].

2.4. Compound Identification

2.4.1. Semi-Preparative HPLC

Semipreparative LC analysis of SEC fractions FR14 and FR18 was performed on an Agilent 1260 Infinity II LC System (Santa Clara, CA, USA) equipped with a quaternary pump, multicolumn thermostat, autosampler and multiple wavelength detector. Separations were performed on an Agilent ZORBAX StableBond C18 (5 µm, 9.4 × 150 mm) at room temperature. SEC fraction FR14 was dissolved in 3 mL 30% methanol to a final concentration of 42 mg/mL. SEC fraction FR18 was dissolved in 3 mL 30% methanol to a final concentration of 15 mg/mL. The mobile phase consisted of water (A) and Acetonitrile (B). The flow rate was 4 mL/min. Signal was detected at 220 nm. Increasing amount of SEC fractions were injected starting from 40 µL, 60 µL, 80 µL, 100 µL, 120 µL, 140 µL and 160 µL for FR14 and 24 µL, 48 µL, 100 µL and 250 µL for FR 18 until all the amount was injected. Individual compounds were manually collected in 50 mL tubes, evaporated to dryness using Labconco CentiVap vacuum concentrator (Kansas city, USA) and used for ^{1}H NMR and LC-MS/MS analysis.

2.4.2. ^{1}H NMR

Dried semi-preparative HPLC fractions were dissolved in 600 µL of CD3OD with deuterated 25 mM 196 potassium phosphate buffer at pH 5.91 in ratio 1:1 (v/v). Proton NMR spectra were acquired on a Bruker NEO 600 spectrometer (600.18 MHz, Biospin GmbH, Rheinstetten, Germany) at 293.0 ± 0.1 K using a Prodigy probehead. Standard parameters have been used—pulse programs zg30 and noesypr1d for experiments with water presaturation, pulse width 30°/90°, spectral width 13.66 ppm, 64 K data points, 1/64 scans, acquisition time 4.0 s, and relaxation delay 4.0 s. The signal of the rest proton signal of the solvent CD3OD at 3.3 ppm was used as an internal reference.

2.4.3. LC-MS/MS

After acquisition of ^{1}H NMR spectra, LC-MS/MS analyses was performed on an Agilent 1260 Infinity II LC System equipped with a quaternary pump, autosampler, multi-column thermostat and Agilent 6546 QTOF detector. Analytical separations were performed on an Agilent InfinityLab Poroshell 120 SB-C18 (2.7 μm, 3 × 150 mm) (Santa Clara, CA, USA) at room temperature. ESI-MS spectra were recorded in negative ion mode between m/z 20–3200. The fragmentor energy of ESI was set to 120 V. The injection volume was 10 μL (1 mg/mL dry weight). The mobile phase consisted of 0.1% aqueous formic acid (A) and 0.1% formic acid in Acetonitrile (B). The flow rate was 0.6 mL/min. The following gradient profile was used for qualitative analysis of SEC fraction FR14: 0 min 15% B, gradient 0–20 min 18% B, 20–22 min 100% B, 22–30 min 100% B isocratic, 30–32 min to 15% B. The following gradient profile was used for qualitative analysis of SEC fraction FR18: 0 min 25% B, 0–6 min isocratic 25% B, gradient 6–12 min 42% B, 12–14 min 100% B, 14–24 min 100% B isocratic, 24–26 min to 25% B. Three different collision-induced dissociation (CID) energies including 10, 20 and 40 eV were used for MS/MS verification of the myconoside structure. Hyspiduline hispidulin 8-C-(6-O-acetyl-2″-O-syringoyl-β-glucopyranoside) was identified using a CID energy of 10 eV during MS/MS analysis.

2.5. Molecular Docking

For the molecular modeling we used MOL structure file of myconoside (J796.651B, Japan Chemical Substance Dictionary (Nikkaji)) as ligand and three receptor macromolecules: human glucose transporter pdb ID 4PYP [36], MYST acetyltransferase pdb ID 6OIO [37] and Estrogen Receptor pdb ID 3OS8 [38]. Complex X-ray structures, including a receptor protein bound to a low-molecular-weight ligand, were used to determine the macromolecular complexes. Prior to docking, the primary ligands were removed. 3D macromolecular docking was performed with the Seamdock [39]. For human glucose transporter Charmm-Gui membrane builder [40] has been used to orientate the transmembrane spans through lipid bilayer and the resulting supramolecular structure was used to bind to the myconoside. All structures were optimized by free energy minimum and visualized with molecular dynamics programs Chimera 1.15 [41] and protein modeling—RasTop.

3. Results

3.1. Cell Viability and Cell Proliferation

3.1.1. Cell Viability after Treatment with *Haberlea rhodopensis* Extracts and Fractions

Our preliminary experiments showed that the application of total extracts to control and breast cancer cell lines for 24 h were not informative enough. Longer exposure (for 48 h) of the cell lines to concentrations of extracts up to 300 μg/mL brought no differences in reaction to the treatment (Table S1). This triggered the application of non-target fractionation of the extracts with liquid chromatography. We obtained 21 fractions and based on the availability of a sufficient amount of dry substance, 11 were selected for further analyses, and performed with 300 μg/mL for 48 h (Table S1). The results obtained were subjected to an ANOVA (Figure 1). A significant reduction in cancer cells viability was achieved after treatment with several of the fractions, among which 14 and 18 were most effective.

To further evaluate the effect of the selected fractions 14 and 18, the treatments have been prolonged for 72 h (Figure 2). The viability of the MCF-10A normal cell line was only slightly reduced to about 80% for both fractions. The viability of MCF7 cell line was significantly reduced below 50% for both fractions, while MDA-MB231 cell line was slightly less affected to 55% ± 10 after a treatment with fraction 14 and to 68% ± 2.2 for fraction 18

(Figure 2). The results obtained gave a good background for further proliferation assays with both fractions.

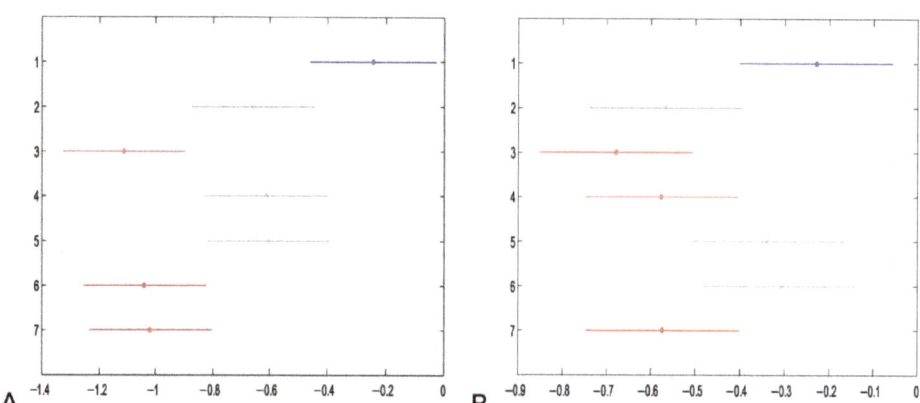

Figure 1. ANOVA multiple comparison plots of two breast cancer cell lines for statistical significance of difference between control conditions and treatment with total extract and different fractions for 48 h. (**A**) MCF7 and (**B**) MDA-MB231. Bars represent the viability of cells shown on x axis as log2 values of measured extinction for each treatment (shown on y axis); 1 untreated; 2—total extract; 3—fr 14; 4—fr 15; 5—fr 16; 6—fr 17; 7—fr 18. The group tested for significance is represented with blue, with red are assigned groups with significant difference from the tested group, and with grey are shown the groups without significant changes.

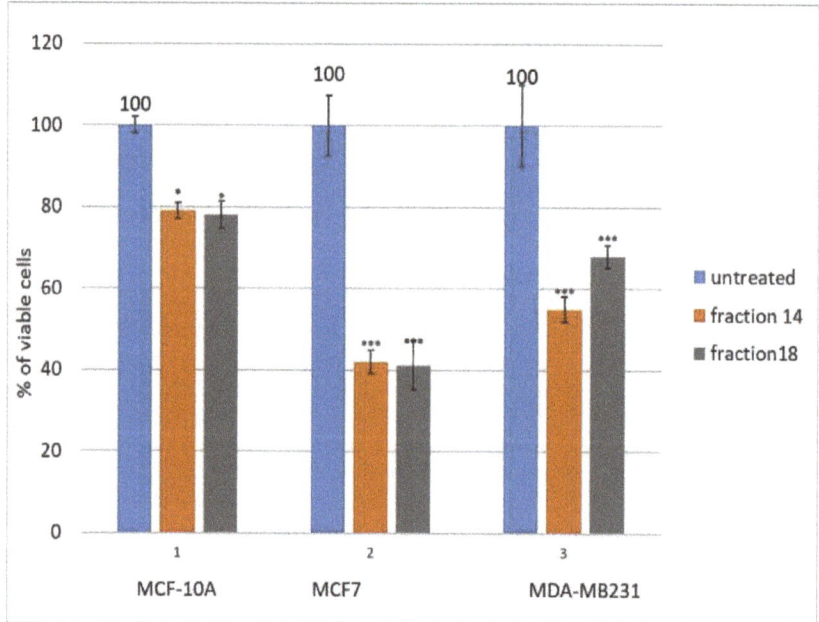

Figure 2. Cell viability of normal breast cell line MCF-10A and breast cancer cell lines MCF7 and MDA-MB231 after treatment with fractions 14 and 18 in concentration 300 µg/mL for 72 h. Untreated samples are used as controls. The standard error bars are shown in percent. *—the means are statistically significant at $p \leq 0.05$; ***—the means are statistically significant at $p \leq 0.001$.

3.1.2. Cell Proliferation Assay after Treatment with Two Selected Fractions

The proliferation assay was performed with the studied cell lines—normal MCF-10A and two cancer cell lines MCF7 and MDA-MB231 (Figure 3). The cells were treated with two *H. rhodopensis* fractions—No. 14 and 18 with a concentration of 300 µg/mL dry substance. The viable cells were scored in absolute numbers at different time points (from 24 to 96 h). The cell line MCF-10A (panel A) had a normal growth and the treated cells showed slightly reduced numbers—reaching about 82.5% of the untreated cells for fraction 14 and 75% for fraction 18 at the end of the assay. The proliferation curves of control and treated cells of line MCF7 were very different (panel B). The numbers of treated cells were reduced at each time point. This was particularly true at the end of the assay where the absolute numbers of proliferation were reduced to 37.2% (fraction 14) and 36.3% (fraction 18). The cell proliferation of triple negative cell line MDA-MB231 was also significantly reduced after treatment with the fractions (Panel C). The reduction started after 24 h treatment and continued till the end of the assay. On the other hand, the proliferation rate of the cells at the last stages of treatment—72 and 96 h formed a plateau. Nevertheless, at the end of the assay, the reduction was 40–45% in comparison with control untreated cells. It should be underlined that in both cancer lines the reduction in proliferation rate showed no differences between both fractions after 48 h of treatment. The proliferation of MCF7 cell line was slightly more reduced than that of the MDA-MB231.

3.2. Identification of Phytoactive Compounds in Plant Extract

To identify the most abundant compounds from the fractions with the strongest effects on cancer lines—Fr 14 and Fr 18 (Figure 4A), we used semi-preparative HPLC to collect the most abundant peaks for each fraction followed by ^{1}H NMR and mass spectrometry for identification (Figures 4B and 5). According to results from MS, the abundances of compounds **1** and **2** represent 24% and 14% from TIC of fractions 18 and 14, respectively (Tables S2 and S3, Figure 4A).

Semi-preparative fraction of the most abundant compound (**2**) from fraction 14 consists exclusively of myconoside as indicated by the NMR spectra (Table S4, Figure 5A). The MS and MS/MS spectra confirmed the mass of the pseudomolecular ion of myconoside (743.2399 (M − H)) as well as the presence of characteristic products of its fragmentation (Figures 4B and S1). This compound consists 14% from the total metabolite content in Fr14 followed by another unidentified compound with 6% of TIC. All other detected compounds showed very low abundances in fraction (Table S2, Figure 4A). The yield of the purified myconoside by semi-preparative fractionation was 34 mg (Figure 5A). Fr 18 shows several compounds above 6% of TIC including myconoside with 8.5% of TIC (Table S3, Figure 4B). The NMR spectra of semi-preparative fraction corresponding to the most abundant peak (23.5% of TIC) showed two main components in ratio 1.7:1, the higher concentrated one corresponding to hispidulin 8-C-(6-O-acetyl-2-O-syringoyl-β-glucopyranoside) (Table S3, Figure 5B). All signal assignments are in line with published data [42,43]. This identification was further confirmed by the corresponding pseudomolecular ion (683.1635 (M − H)) and fragmentation products from the MS^2 spectra corresponding to the loss of syringoyl and sugar moiety (Figures 5B and S2). However, further purification steps and analyses are needed for better evaluation of the active compound in this fraction.

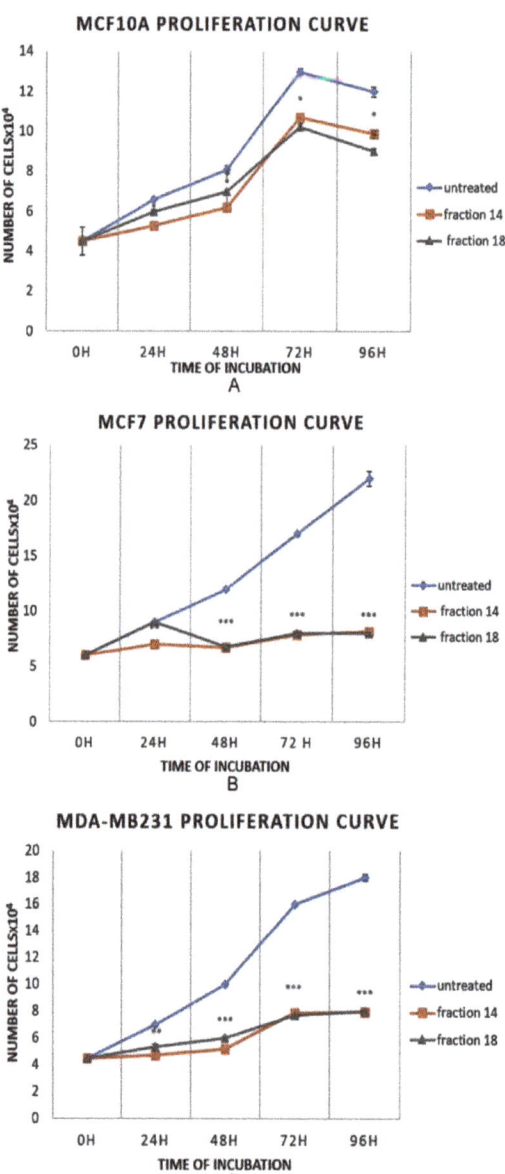

Figure 3. Proliferation curves of normal epithelial breast cell line (**A**) and breast cancer cell lines MCF7 (**B**) and MDA-MB231 (**C**) treated with *H. rhodopensis* fractions N14 and N18. The standard error bars are shown. All means at each time point are statistically significant at $p \leq 0.05$ (*), $p \leq 0.01$ (**), and $p < 0.001$ (***).

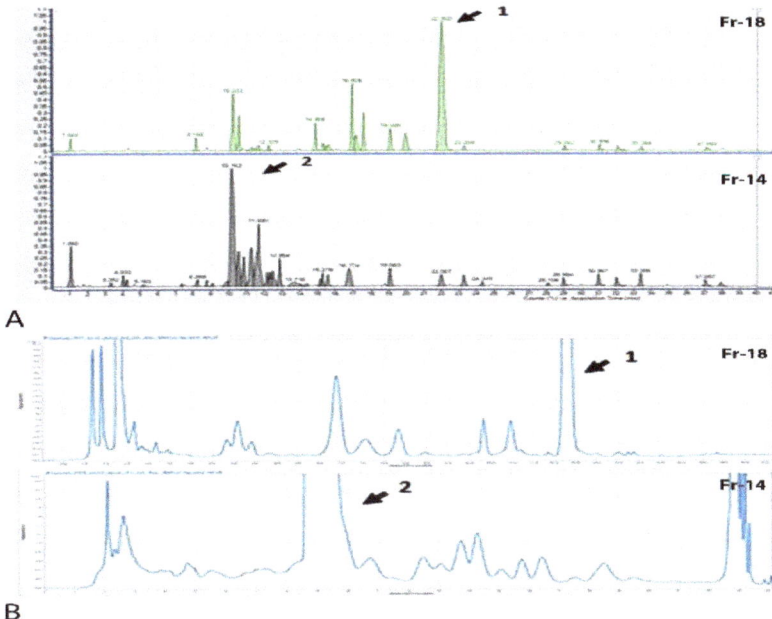

Figure 4. Mass spectrometry analysis of SEC fractions with significant effect on cancer lines and semi preparative fractionation of most abundant compounds. (**A**) Chromatograms of MS spectra of fractions 14 and 18. The most abundant peaks in each fraction are designated with numbers. (**B**) HPLC chromatograms of semi-preparative purification of the most abundant compounds from fraction 14 and 18.

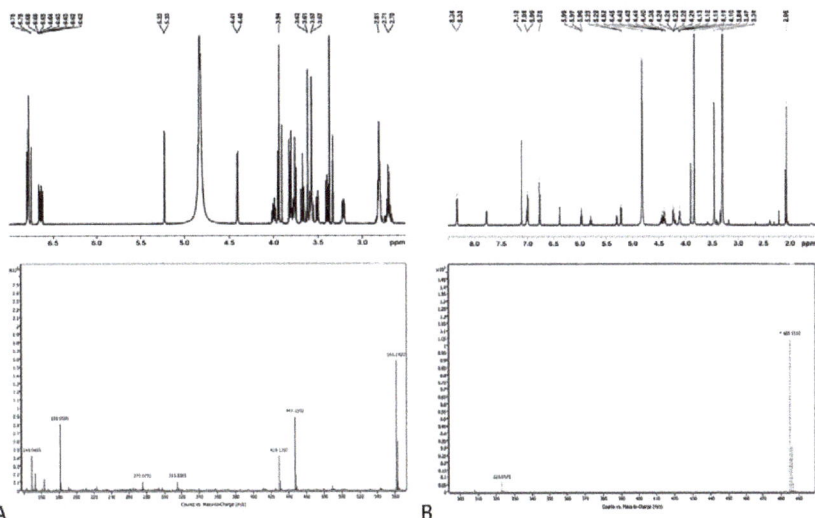

Figure 5. Identification of the most abundant compounds in fractions 14 and 18 by combined NMR/MS2 analysis on same sample. The 1H NMR spectra for each compound is represented according to chemical shift (ppm) (upper panel), while ions of fragmentation from MS2 are represented according to their m/z (lower panel). (**A**) Identification of purified myconoside (compound 2) from fraction 14. (**B**) Identification of purified hispidulin 8-C-(6-O-acetyl-2-O-syringoyl-b-glucopyranoside) from fraction 18.

3.3. Docking Analysis of Myconoside with Breast Cancer Proteins

Considering our ability to purify and identify myconoside as the main compound in Fr 14 which significantly reduced cancer cell viability and proliferation and the available 2D and 3D deposited structures, we performed flexible docking analysis of this glycoside with several breast cancer proteins involved in cellular transport, signaling and DNA modification (Figure 6A).

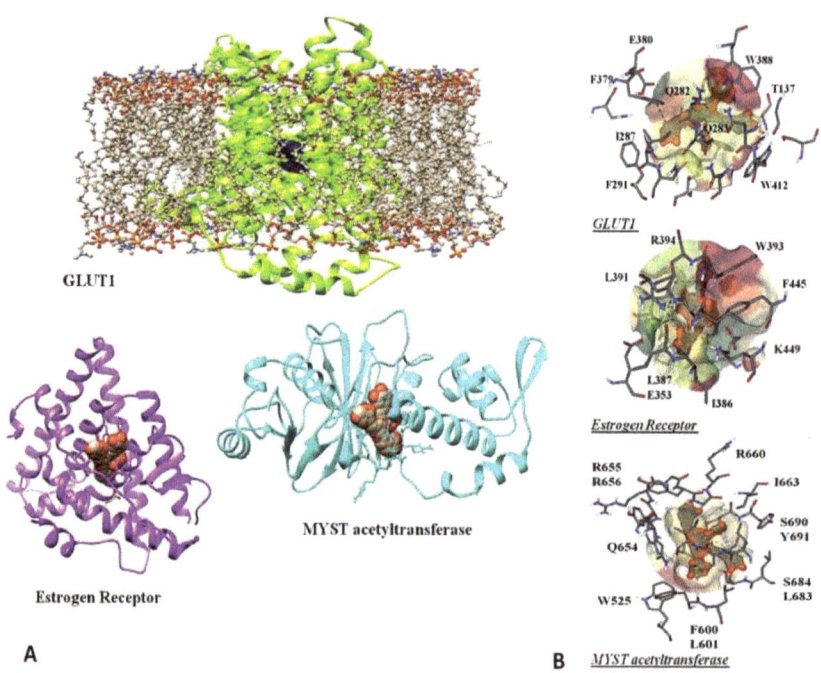

Figure 6. Molecular docking of myconoside with several protein targets from breast cancer lines. (**A**) Ribbons of three-dimensional structure of binding of myconoside with GLUT 1, MYST acetyltransferase and estrogen receptor. The protein backbone is represented as a cartoon with different colors for each protein. The docked myconoside is represented with 3D stick model of chemical formula. GLUT 1 transporter (green) is integrated and oriented in membrane phospholipid bilayer; myconoside is given in blue. (**B**) Docking of myconoside into the binding cavity of the proteins with the corresponding intermolecular interactions and amino acid residues.

Our results showed that myconoside can interact with the three tested proteins. However, the interaction with GLUT 1 displayed more binding affinity according to ΔG (−19.8 kcal/mol) than other two proteins—MYST acetyltransferase (−12.3 kcal/mol) and Estrogen Receptor (−4.2 kcal/mol). Residues of the amino acids E380, F379, F287, F291, W412, T137 and W388 from GLUT 1 transporter are involved in interactions with myconoside. The amino acid residues E353, I386, L387, L391, W393, R394, F445 and K449 of estrogen receptor are involved in the represented receptor-ligand complex; while R655, R656, R660, F663, S690, Y691, S684, L683, F600, L601, W525 and Q654 were assigned in binding pocket of MYST acetyltransferase (Figure 6B). The docking model of myconoside with GLUT 1 is mainly determined by hydrogen bonds and hydrophobic interactions of Phenylalanine and Tryptophan residues. Most interactions with estrogen receptor are in a hydrophobic manner namely by Phenylalanine and Tryptophan residues as well by Leucine and Isoleucine residues. The binding of myconoside with MYST acetyltransferase is due to hydrophobic interactions of Leucine and Isoleucine residues as well as hydrogen bonds with Arginine and Serine.

4. Discussion

Breast cancer is among the most challenging human diseases. Despite the significant progress achieved in cancer treatment, the search for new natural products continues to be very intensive. Plant metabolites are tested for possible anticancer effects. Some of them can be used as food additives for cancer prevention or as therapeutics of side effects relief after radiotherapy. Others are used to enhance the effect of conventional drugs [11,12,44,45]. In this respect, promising results were reported for the proliferation rate reduction and cytostatic effects of some plant-derived alkaloids [10,33,46]; however, their application was limited by the multi-drug resistance developed by the cancer cells. The additional burden of the side effects of chemo- or radio-therapy paves the way for further studies on new potential sources of useful compounds.

Resurrection plants are a rich source of secondary metabolites with high antioxidant potential. Here, we describe, we believe for the first time, promising results of breast cancer cell lines treatment with extracts and fractions derived from leaves of the Balkan resurrection plant species *Haberlea rhodopensis*. The application of total leaf extracts, obtained with various extraction agents led to encouraging results in studies with several human diseases, including some types of cancer [27–30]. Viability reduction was described in two prostate cancer cell lines after methanol extracts treatment [27]. The same types of extracts were reported to have unique synergetic inhibitory effects against the herpes virus [29]. In addition, they could be a good candidate to be involved in complex treatments of pathological dermatological conditions [28]. Recently [30] extensive study with six human cancer cell lines—A549 (non-small cell lung adenocarcinoma, HepG2 (hepatocellular carcinoma)), HT29 and Caco-2 (colorectal adenocarcinomas), and PC3 and DU145 (prostate adenocarcinomas) showed that alcohol extraction appeared to be more effective than the aqueous. Significant antimigratory concentration-dependent effects were achieved for non-small cell lung adenocarcinoma and hepatocellular carcinoma (HepG2) cell lines [30].

Our study showed that the total extract was unable to significantly reduce the viability of the MCF7 and MDA-MB231 breast cancer cell lines (Supplementary Table S1). This triggered our further efforts to fractionate the extracts to achieve enrichment of active substances. Two of the fractions (14 and 18) showed significant effects on breast cancer cell viability (Figure 1) and a negligible effect on normal breast epithelia cell line MCF10-A (Supplementary Table S1, Figure 2). These fractions had a high impact on cell proliferation of the studied breast cancer cell lines and an insignificant effect on the proliferation of the normal breast epithelial cell line. One of the fractions (Fr 14) is enriched of myconoside and another one (Fr 18)—of hispidulin 8-C-(6-O-acetyl-2″-O-syringoyl-β-glu-copyranoside) (Figure 4).

4.1. Potential Role of Myconoside

Myconoside has been isolated earlier from the three European members of Gesnereiacea, including *H. rhodopensis* [42,47–51]. The myconoside molecule has a phenyl glycoside structure with 12 hydroxyl groups (Supplementary Figure S1) which determined its chemical activity and possibility to form a 3D structure which was deposited in the Japan Chemical Substance Dictionary (Nikkaji) database. The structure file was used to propose models of its binding to three proteins with an essential role for breast cancer and breast cancer cell lines development—estrogen receptor, glucose transporter GLUT1 and MYST acetyltrasferase (Figure 6). All of these three proteins are expressed in the MCF7 cell line whereas two of them—GLUT1 and MYST acetyltrasferase are expressed in MB-MDA231. We presumed a possibility for myconoside binding to glucose transporter GLUT1. The GLUT family of transporters are localized on the cell membrane and are connected by hydrophilic loops [52]. They are expressed in high levels in different types of tumors, including breast tumors [53–55] and in particular in the two cancer cell lines under study—MCF7 and MDA-MB231 [6]. The chemically synthesized transporter inhibitors WZB117 and STF-31 block cell proliferation of MCF7 and MDA-MB231 cell lines by an increase in extracellular glucose and a decrease in extracellular lactose [55]. WZB27 and

WZB115 are two transporter inhibitors which reduce glucose uptake and block cell proliferation in MCF7 cell line [56]. Various polyphenol substances of plant origin—gossypol, genistein, resveratrol, quercetin have been described to influence glucose metabolism in breast cancer cell lines [7]. We have two hypotheses about myconoside binding to the glucose transporter. It could be a glucose transporter blocker, thus reducing glucose uptake in cells, or this could be the way for myconoside penetration in the cell. The prediction of molecular binding of myconoside with GLUT1 transporter makes possible the penetration of myconoside in cancer cells by binding to this membrane localized transporter. These hypotheses should be a subject of future studies.

The estrogen receptor has an essential role for estrogen-dependent growth in estrogen expressing tumors and cancer cell lines. We presume a possible binding of myconoside to the estrogen receptor of the MCF7 cell line—a good model to search possibilities for estrogen receptor agonists/antagonists in breast cancer therapy [57]. Drugs with polyphenyl structures block the DNA binding receptor domain and are competitive antagonists of estradiol. This is a mechanism for hormone dependent growth blockage of breast cancer cells [57,58]. Such estrogen antagonists of plant origin are the coumarins with antiproliferative effects on the breast cancer cell line MDA-MB435 [57]. We propose a role of myconoside as an estrogen receptor modulator. Plant substances were virtually screened and 162 of them have been validated by docking with estrogen receptor α. Eight of them have ER α competition effects. Genistein, daidzein, phloretin, ellagic acid, ursolic acid, (−)-epigallocatechin-3-gallate, kaempferol are with different antagonistic activities against estrogen receptor α [59].

Our docking analyses allowed the presumption that another target of myconoside could be MYST acetyltransferase. This could be a mechanism, affecting cell proliferation in studied breast cell lines related to epigenetic DNA regulation. We found that interactions of MYST acetyltransferase with myconoside are docked by amino acids residues ARG655, LEU686, GLN760, ARG660, LEU689 and LYS763, which are previously reported to interact with different compounds of the medicinal plant *Withania somnifera* (L.) [60]. The MYST acetiytransferases are related to the activation of estrogen receptor α by acetylation of the hystone acetyltransferase domain in the estrogen receptor promotor. This mechanism of epigenetic regulation activates the estrogen receptor expression in estrogen receptor positive breast cancer cell lines like MCF7 [7]. The blockage of acetyltransferase enzyme suppresses the activation of estrogen receptor α promotor. Studies of acetyltransferase mRNA and protein expression showed different levels of their expression in a panel of breast cancer cell lines [7]. This is a pathway to limit estrogen dependent growth in estrogen receptor positive cells and could be a possible explanation for the higher inhibitory effect on the triple positive cell line MCF7. Purified myconoside was shown to have antimygratory effects and cell proliferation blockage on the A549 lung adenocarcinoma cell line with an IC50 of 20 μg/mL [61]. Our results showed that fractions containing myconoside did not significantly affect cell viability or proliferation of the normal cell line MCF10A (Figures 2 and 3). This is in agreement with the report that *H. rhodopensis* extracts fractions with identified myconoside and calceolarioside E has effects on protein expression of neutrophil essential transcription factor regulating redox potential Nrf2 in bone marrow neutrophils [42]. In addition, when applied in low concentrations, the natural phenyl glycoside induces hormetic-like response in MDCKII cell line [62] or has photoprotective effects on UVA/UVB irradiated immortalized keratinocytes [53].

4.2. Potential Role of Hispidulin 8-C-(6-O-acetyl-2″-O-syringoyl-β-glu-copyranoside)

The proliferation rate was significantly reduced in both breast cancer cell lines after treatment with Fraction 18. This fraction is more complex, containing several compounds, including myconoside and hispidulin 8-C-(6-O-acetyl-2″-O-syringoyl-β-glu-copyranoside). We identified the most abundant compound as hispidulin 8-C-(6-O-acetyl-2″-O-syringoyl-β-glu-copyranoside) (Figure 5B). It was reported earlier in a mixture of flavone glycosides of the same plant species [43,63]. While myconoside is found exceptionally in some

Gesmneriaceae resurrection plant species, hispidulin is common for many plant species widely applicable in traditional medicine, such as *Grindelia argentina*, *Arrabidaea chica*, *Saussurea involucrate*, *Crossostephium chinense*, *Artemisia* and *Salvia* species. It was shown to possess various activities—antioxidant, antifungal, anti-inflammatory, antimutagenic, and antitumor [64]. The potential therapeutic usefulness of hispidulin has been studied in a variety of tumors [65–68]. The molecular mechanisms mediating its effects on cancer cell lines of different origin have been analyzed. Effects on cell viability and proliferation have been established for prostate cancer cell lines [65], GLUT1-HEK293T transformed cells and Hep2G hepatocellular carcinoma cells [66], as well as for human melanoma A253 cancer cells [67]. Its mechanism of influence on cell signal pathways depends on the type of cancer. In the case of prostate cancer cell lines it is related to limitation of cell migration, invasion, proliferation and apoptosis initiation mediated via PPARγ activation and autophagy induction [65]. Hispidulin modulates epithelial-mesenchymal transition in breast cancer cells—a process associated with the disruption of cell junctions, increase in cell mobility and metastasis by suppressing the TGF-β1-induced Smad2/3 signaling pathway [68]. The effects on human melanoma cells are mediated by activation of apoptosis rather than autophagy, inhibiting kinase signaling pathways AKT and ERK [4]. Recently, the inhibitory effect of various flavonoids on GLUT1 transporter in HEK293T and HepG2 has been reported [66]. Several flavonoids, including hispidulin inhibit hepatocellular carcinoma cell line Hep2G and GLUT1 expressing HEK2893T cell line by binding to glucose transporter1, which was shown by docking analysis and validated [66]. The suppression of GLUT1 transporter activity by hispidulin was established to 40% of control untreated sample at the concentration range 100–150 μM. This fact could explain our results on cell viability reduction and suppression of proliferation in the studied cell line MCF7 and MDA-MB231, which both express GLUT1 in high levels and are characterized with intensive glucose metabolism. The combination of identified hispidulin 8-C-(6-O-acetyl-2″-O-syringoyl-β-glu- copyranoside) and the second highly abundant unidentified substance in the fraction should be a subject of future investigations related to its identification and the clarification of molecular mechanisms of influence on breast cancer cell proliferation.

5. Conclusions

This study described for the first time the effect of *H. rhodopensis* methanol extract fractions on the viability and proliferation of two breast cancer cell lines with different characteristics and a normal cell line. The inhibitory effects are specific for cancer cell lines and are better for the hormone responsive one. Myconoside appears to be a suitable agent for cancer therapy and a possible model for its action was proposed, targeting three specific cancer hallmark proteins.

Supplementary Materials: The following supporting information can be downloaded at: https://www.mdpi.com/article/10.3390/separations10070388/s1. Table S1. cell viability, Table S2. Metabolic content of fraction 14, Table S3. Metabolic content of fraction 18, Table S4. signals of ^{1}H and ^{13}C of myconoside, Table S5. ^{1}H signals of hispidulin, Figure S1. Molecular structure of myconoside (MS/MS identification), Figure S2. Molecular structure of hispidulin (MS/MS identification), Figure S3. Structure of myconoside (NMR), Figure S4. Structure of hispidulin (NMR).

Author Contributions: Conceptualization, D.Z. and P.M.; methodology, D.Z. and P.M.; validation, D.Z., P.M., K.R. and S.S.; formal analysis, D.Z., P.M., S.S. and K.R.; investigation, D.Z., P.M., S.S., K.R., S.Z., L.S.-S., D.M. and D.D.; resources, D.Z., L.S.-S., D.D., S.S. and K.R.; writing—original draft preparation, D.Z., P.M., D.M., L.S.-S. and D.D.; writing—review and editing, D.Z., P.M., S.S., K.R., S.Z., L.S.-S., D.M. and D.D.; visualization, D.Z., P.M., S.S. and K.R.; supervision, D.Z., P.M., S.S. and D.D.; project administration, D.Z., P.M., D.D. and L.S.-S.; funding acquisition, D.Z. All authors have read and agreed to the published version of the manuscript.

Funding: The research was funded by NSF of Bulgaria, Grant number КП-06-Н41/6, Operational Program Science and Education for Smart Growth 2014–2020, co-financed by the European Union through the European Structural and Investment Funds, Grant BG05M2OP001-1.002-0012.

Data Availability Statement: Data is contained within the article.

Acknowledgments: The authors highly appreciate the help of Zlatina Gospodinova (Institute of Plant Physiology and Genetics, Bulgarian Academy of Sciences) by kindly providing cell line MCF-10A and of Radostina Alexandrova (Institute of Experimental Morphology and Anthropology Bulgarian Academy of Sciences, Bulgarian Academy of Sciences) for kindly providing MCF7 cell line and of Milena Mourdjeva (Institute of Biology and Immunology of Reproduction, Bulgarian Academy of Sciences) for kindly providing cell line MDA-MB231. We highly appreciate the help of Svetlana Hristova, Ph.D (Department of Medical Physics and Biophysics, Medical Faculty, Medical University–Sofia, Zdrave Str. 2, 1431 Sofia, Bulgaria) for kindly contribute with software for analyses of molecular docking.

Conflicts of Interest: The authors declare no conflict of interest.

References

1. Lancet, T. Breast Cancer in Developing Countries. *Lancet* **2009**, *374*, 1567. [CrossRef]
2. Arnold, M.; Morgan, E.; Rumgay, H.; Mafra, A.; Singh, D.; Laversanne, M.; Vignat, J.; Gralow, J.R.; Cardoso, F.; Siesling, S.; et al. Current and Future Burden of Breast Cancer: Global Statistics for 2020 and 2040. *Breast* **2022**, *66*, 15–23. [CrossRef]
3. Breast Cancer Facts & Statistics 2023. Available online: https://www.breastcancer.org/facts-statistics (accessed on 27 March 2023).
4. Shrihastini, V.; Muthuramalingam, P.; Adarshan, S.; Sujitha, M.; Chen, J.-T.; Shin, H.; Ramesh, M. Plant Derived Bioactive Compounds, Their Anti-Cancer Effects and In Silico Approaches as an Alternative Target Treatment Strategy for Breast Cancer: An Updated Overview. *Cancers* **2021**, *13*, 6222. [CrossRef] [PubMed]
5. George, B.P.A.; Abrahamse, H. A Review on Novel Breast Cancer Therapies: Photodynamic Therapy and Plant Derived Agent Induced Cell Death Mechanisms. *Anticancer Agents Med. Chem.* **2016**, *16*, 793–801. [CrossRef]
6. Barbosa, A.M.; Martel, F. Targeting Glucose Transporters for Breast Cancer Therapy: The Effect of Natural and Synthetic Compounds. *Cancers* **2020**, *12*, 154. [CrossRef] [PubMed]
7. Yu, L.; Liang, Y.; Cao, X.; Wang, X.; Gao, H.; Lin, S.-Y.; Schiff, R.; Wang, X.-S.; Li, K. Identification of MYST3 as a Novel Epigenetic Activator of ERα Frequently Amplified in Breast Cancer. *Oncogene* **2017**, *36*, 2910–2918. [CrossRef] [PubMed]
8. Ali Abdalla, Y.O.; Subramaniam, B.; Nyamathulla, S.; Shamsuddin, N.; Arshad, N.M.; Mun, K.S.; Awang, K.; Nagoor, N.H. Natural Products for Cancer Therapy: A Review of Their Mechanism of Actions and Toxicity in the Past Decade. *J. Trop. Med.* **2022**, *2022*, e5794350. [CrossRef] [PubMed]
9. Lopes, C.M.; Dourado, A.; Oliveira, R. Phytotherapy and Nutritional Supplements on Breast Cancer. *BioMed Res. Int.* **2017**, *2017*, e7207983. [CrossRef]
10. Mazurakova, A.; Koklesova, L.; Samec, M.; Kudela, E.; Kajo, K.; Skuciova, V.; Csizmár, S.H.; Mestanova, V.; Pec, M.; Adamkov, M.; et al. Anti-Breast Cancer Effects of Phytochemicals: Primary, Secondary, and Tertiary Care. *EPMA J.* **2022**, *13*, 315–334. [CrossRef]
11. Küpeli Akkol, E.; Genç, Y.; Karpuz, B.; Sobarzo-Sánchez, E.; Capasso, R. Coumarins and Coumarin-Related Compounds in Pharmacotherapy of Cancer. *Cancers* **2020**, *12*, 1959. [CrossRef]
12. Li, Z.; Li, J.; Mo, B.; Hu, C.; Liu, H.; Qi, H.; Wang, X.; Xu, J. Genistein Induces Cell Apoptosis in MDA-MB-231 Breast Cancer Cells via the Mitogen-Activated Protein Kinase Pathway. *Toxicol. In Vitro* **2008**, *22*, 1749–1753. [CrossRef] [PubMed]
13. Melo, J.O.; Fachin, A.L.; Rizo, W.F.; Jesus, H.C.R.; Arrigoni-Blank, M.F.; Alves, P.B.; Marins, M.A.; França, S.C.; Blank, A.F. Cytotoxic Effects of Essential Oils from Three Lippia Gracilis Schauer Genotypes on HeLa, B16, and MCF-7 Cells and Normal Human Fibroblasts. *Genet. Mol. Res.* **2014**, *13*, 2691–2697. [CrossRef]
14. Woo, C.C.; Hsu, A.; Kumar, A.P.; Sethi, G.; Tan, K.H.B. Thymoquinone Inhibits Tumor Growth and Induces Apoptosis in a Breast Cancer Xenograft Mouse Model: The Role of P38 MAPK and ROS. *PLoS ONE* **2013**, *8*, e75356. [CrossRef] [PubMed]
15. Borges, R.M.; Resende, J.V.M.; Pinto, A.P.; Garrido, B.C. Exploring Correlations between MS and NMR for Compound Identification Using Essential Oils: A Pilot Study. *Phytochem. Anal.* **2022**, *33*, 533–542. [CrossRef]
16. Leggett, A.; Wang, C.; Li, D.-W.; Somogyi, A.; Bruschweiler-Li, L.; Brüschweiler, R. Identification of Unknown Metabolomics Mixture Compounds by Combining NMR, MS, and Cheminformatics. *Methods Enzymol.* **2019**, *615*, 407–422. [CrossRef] [PubMed]
17. Gaff, D.F.; Oliver, M. The Evolution of Desiccation Tolerance in Angiosperm Plants: A Rare yet Common Phenomenon. *Funct Plant Biol.* **2013**, *40*, 315–328. [CrossRef]
18. Dinakar, C.; Bartels, D. Desiccation Tolerance in Resurrection Plants: New Insights from Transcriptome, Proteome and Metabolome Analysis. *Front. Plant Sci.* **2013**, *4*, 482. [CrossRef]
19. Legardón, A.; García-Plazaola, J.I. Gesneriads, a Source of Resurrection and Double-Tolerant Species: Proposal of New Desiccation- and Freezing-Tolerant Plants and Their Physiological Adaptations. *Biology* **2023**, *12*, 107. [CrossRef]
20. Ivanova, A.; O'Leary, B.; Signorelli, S.; Falconet, D.; Moyankova, D.; Whelan, J.; Djilianov, D.; Murcha, M.W. Mitochondrial Activity and Biogenesis during Resurrection of *Haberlea rhodopensis*. *New Phytol.* **2022**, *236*, 943–957. [CrossRef]
21. Liu, J.; Moyankova, D.; Lin, C.-T.; Mladenov, P.; Sun, R.-Z.; Djilianov, D.; Deng, X. Transcriptome Reprogramming during Severe Dehydration Contributes to Physiological and Metabolic Changes in the Resurrection Plant *Haberlea rhodopensis*. *BMC Plant Biol.* **2018**, *18*, 351. [CrossRef]

22. Mladenov, P.; Zasheva, D.; Planchon, S.; Leclercq, C.C.; Falconet, D.; Moyet, L.; Brugière, S.; Moyankova, D.; Tchorbadjieva, M.; Ferro, M.; et al. Proteomics Evidence of a Systemic Response to Desiccation in the Resurrection Plant *Haberlea rhodopensis*. *Int. J. Mol. Sci.* **2022**, *23*, 8520. [CrossRef] [PubMed]
23. Vassileva, V.; Moyankova, D.; Dimitrova, A.; Mladenov, P.; Djilianov, D. Assessment of Leaf Micromorphology after Full Desiccation of Resurrection Plants. *Plant Biosyst.* **2019**, *153*, 108–117. [CrossRef]
24. Mladenov, P.; Finazzi, G.; Bligny, R.; Moyankova, D.; Zasheva, D.; Boisson, A.-M.; Brugière, S.; Krasteva, V.; Alipieva, K.; Simova, S.; et al. In Vivo Spectroscopy and NMR Metabolite Fingerprinting Approaches to Connect the Dynamics of Photosynthetic and Metabolic Phenotypes in Resurrection Plant *Haberlea rhodopensis* during Desiccation and Recovery. *Front. Plant Sci.* **2015**, *6*, 564. [CrossRef] [PubMed]
25. Kuroki, S.; Tsenkova, R.; Moyankova, D.; Muncan, J.; Morita, H.; Atanassova, S.; Djilianov, D. Water Molecular Structure Underpins Extreme Desiccation Tolerance of the Resurrection Plant *Haberlea rhodopensis*. *Sci. Rep.* **2019**, *9*, 3049. [CrossRef] [PubMed]
26. Djilianov, D.; Ivanov, S.; Georgieva, T.; Moyankova, D.; Berkov, S.; Petrova, G.; Mladenov, P.; Christov, N.; Hristozova, N.; Peshev, D.; et al. A Holistic Approach to Resurrection Plants. *Haberlea rhodopensis* —A Case Study. *Biotechnol. Biotechnol. Equip.* **2009**, *23*, 1414–1416. [CrossRef]
27. Hayrabedyan, S.; Todorova, K.; Zasheva, D.; Moyankova, D.; Georgieva, D.; Todorova, J.; Djilianov, D. *Haberlea rhodopensis* Has Potential as a New Drug Source Based on Its Broad Biological Modalities. *Biotechnol. Biotechnol. Equip.* **2013**, *27*, 3553–3560. [CrossRef]
28. Kostadinova, A.; Doumanov, J.; Moyankova, D.; Ivanov, S.; Mladenova, K.; Djilianov, D.; Topuzova-Hristova, T. *Haberlea rhodopensis* Extracts Affect Cell Periphery of Keratinocytes. *Comptes Rendus Acad. Bulg. Sci.* **2016**, *69*, 439–448.
29. Moyankova, D.; Hinkov, A.; Shishkov, S.; Djilianov, D. Inhibitory Effect of Extracts from *Haberlea rhodopensis* Friv. against Herpes Simplex Virus. *Comptes Rendus Acad. Bulg. Sci.* **2014**, *76*, 1369–1376.
30. Spyridopoulou, K.; Kyriakou, S.; Nomikou, A.; Roupas, A.; Ermogenous, A.; Karamanoli, K.; Moyankova, D.; Djilianov, D.; Galanis, A.; Panayiotidis, M.I.; et al. Chemical Profiling, Antiproliferative and Antimigratory Capacity of *Haberlea rhodopensis* Extracts in an In Vitro Platform of Various Human Cancer Cell Lines. *Antioxidants* **2022**, *11*, 2305. [CrossRef]
31. Djilianov, D.; Genova, G.; Parvanova, D.; Zapryanova, N.; Konstantinova, T.; Atanassov, A. In Vitro Culture of the Resurrection Plant *Haberlea rhodopensis*. *Plant Cell Tiss. Organ Cult.* **2005**, *80*, 115–118. [CrossRef]
32. Mosmann, T. Rapid Colorimetric Assay for Cellular Growth and Survival: Application to Proliferation and Cytotoxicity Assays. *J. Immunol. Methods* **1983**, *65*, 55–63. [CrossRef]
33. Soprano, M.; Sorriento, D.; Rusciano, M.R.; Maione, A.S.; Limite, G.; Forestieri, P.; D'Angelo, D.; D'Alessio, M.; Campiglia, P.; Formisano, P.; et al. Oxidative Stress Mediates the Antiproliferative Effects of Nelfinavir in Breast Cancer Cells. *PLoS ONE* **2016**, *11*, e0155970. [CrossRef]
34. Standard Error Calculator (High Precision). Available online: https://miniwebtool.com/standard-error-calculator/ (accessed on 27 March 2023).
35. One-Way ANOVA Calculator, Plus Tukey HSD. Available online: https://www.socscistatistics.com/tests/anova/default2.aspx (accessed on 27 March 2023).
36. Deng, D.; Xu, C.; Sun, P.; Wu, J.; Yan, C.; Hu, M.; Yan, N. Crystal Structure of the Human Glucose Transporter GLUT1. *Nature* **2014**, *510*, 121–125. [CrossRef] [PubMed]
37. Leaver, D.J.; Cleary, B.; Nguyen, N.; Priebbenow, D.L.; Lagiakos, H.R.; Sanchez, J.; Xue, L.; Huang, F.; Sun, Y.; Mujumdar, P.; et al. Discovery of Benzoylsulfonohydrazides as Potent Inhibitors of the Histone Acetyltransferase KAT6A. *J. Med. Chem.* **2019**, *62*, 7146–7159. [CrossRef]
38. Bruning, J.B.; Parent, A.A.; Gil, G.; Zhao, M.; Nowak, J.; Pace, M.C.; Smith, C.L.; Afonine, P.V.; Adams, P.D.; Katzenellenbogen, J.A.; et al. Coupling of Receptor Conformation and Ligand Orientation Determine Graded Activity. *Nat. Chem. Biol.* **2010**, *6*, 837–843. [CrossRef]
39. Morris, G.M.; Huey, R.; Lindstrom, W.; Sanner, M.F.; Belew, R.K.; Goodsell, D.S.; Olson, A.J. AutoDock4 and AutoDockTools4: Automated Docking with Selective Receptor Flexibility. *J. Comput. Chem.* **2009**, *30*, 2785–2791. [CrossRef] [PubMed]
40. Wu, E.L.; Cheng, X.; Jo, S.; Rui, H.; Song, K.C.; Dávila-Contreras, E.M.; Qi, Y.; Lee, J.; Monje-Galvan, V.; Venable, R.M.; et al. CHARMM-GUI Membrane Builder toward Realistic Biological Membrane Simulations. *J. Comput. Chem.* **2014**, *35*, 1997–2004. [CrossRef] [PubMed]
41. Pettersen, E.F.; Goddard, T.D.; Huang, C.C.; Couch, G.S.; Greenblatt, D.M.; Meng, E.C.; Ferrin, T.E. UCSF Chimera—A Visualization System for Exploratory Research and Analysis. *J. Comput. Chem.* **2004**, *25*, 1605–1612. [CrossRef] [PubMed]
42. Amirova, K.M.; Dimitrova, P.A.; Marchev, A.S.; Krustanova, S.V.; Simova, S.D.; Alipieva, K.I.; Georgiev, M.I. Biotechnologically-Produced Myconoside and Calceolarioside E Induce Nrf2 Expression in Neutrophils. *Int. J. Mol. Sci.* **2021**, *22*, 1759. [CrossRef]
43. Ebrahimi, S.N.; Gafner, F.; Dell'Acqua, G.; Schweikert, K.; Hamburger, M. Flavone 8-C-Glycosides from *Haberlea rhodopensis* Friv. (Gesneriaceae). *Helv. Chim. Acta* **2011**, *94*, 38–45. [CrossRef]
44. Brenton, J.D.; Carey, L.A.; Ahmed, A.A.; Caldas, C. Molecular Classification and Molecular Forecasting of Breast Cancer: Ready for Clinical Application? *J. Clin. Oncol.* **2005**, *23*, 7350–7360. [CrossRef] [PubMed]
45. Demain, A.L.; Vaishnav, P. Natural Products for Cancer Chemotherapy. *Microb. Biotechnol.* **2011**, *4*, 687–699. [CrossRef]

46. Mazumder, K.; Biswas, B.; Raja, I.M.; Fukase, K. A Review of Cytotoxic Plants of the Indian Subcontinent and a Broad-Spectrum Analysis of Their Bioactive Compounds. *Molecules* **2020**, *25*, 1904. [CrossRef] [PubMed]
47. Moyankova, D.; Mladenov, P.; Berkov, S.; Peshev, D.; Georgieva, D.; Djilianov, D. Metabolic Profiling of the Resurrection Plant *Haberlea rhodopensis* during Desiccation and Recovery. *Physiol. Plant.* **2014**, *152*, 675–687. [CrossRef]
48. Cañigueral, S.; Salvía, M.J.; Vila, R.; Iglesias, J.; Virgili, A.; Parella, T. New Polyphenol Glycosides from *Ramonda myconi*. *J. Nat. Prod.* **1996**, *59*, 419–422. [CrossRef]
49. Jensen, S.R. Caffeoyl Phenylethanoid Glycosides in *Sanango racemosum* and in the Gesneriaceae. *Phytochemistry* **1996**, *43*, 777–783. [CrossRef]
50. Kondeva-Burdina, M.; Zheleva-Dimitrova, D.; Nedialkov, P.; Girreser, U.; Mitcheva, M. Cytoprotective and Antioxidant Effects of Phenolic Compounds from *Haberlea rhodopensis* Friv. (Gesneriaceae). *Pharmacogn. Mag.* **2013**, *9*, 294–301. [CrossRef]
51. Gođevac, D.; Ivanović, S.; Simić, K.; Anđelković, B.; Jovanović, Ž.; Rakić, T. Metabolomics Study of the Desiccation and Recovery Process in the Resurrection Plants *Ramonda serbica* and *R. nathaliae*. *Phytochem. Anal.* **2022**, *33*, 961–970. [CrossRef]
52. Mueckler, M.; Thorens, B. The SLC2 (GLUT) Family of Membrane Transporters. *Mol. Asp. Med.* **2013**, *34*, 121–138. [CrossRef]
53. Szablewski, L. Expression of Glucose Transporters in Cancers. *Biochim. Biophys. Acta* **2013**, *1835*, 164–169. [CrossRef]
54. Medina, R.A.; Owen, G.I. Glucose Transporters: Expression, Regulation and Cancer. *Biol. Res.* **2002**, *35*, 9–26. [CrossRef]
55. Xintaropoulou, C.; Ward, C.; Wise, A.; Marston, H.; Turnbull, A.; Langdon, S.P. A Comparative Analysis of Inhibitors of the Glycolysis Pathway in Breast and Ovarian Cancer Cell Line Models. *Oncotarget* **2015**, *6*, 25677–25695. [CrossRef]
56. Liu, Y.; Zhang, W.; Cao, Y.; Liu, Y.; Bergmeier, S.; Chen, X. Small Compound Inhibitors of Basal Glucose Transport Inhibit Cell Proliferation and Induce Apoptosis in Cancer Cells via Glucose-Deprivation-like Mechanisms. *Cancer Lett.* **2010**, *298*, 176–185. [CrossRef] [PubMed]
57. Jameera Begam, A.; Jubie, S.; Nanjan, M.J. Estrogen Receptor Agonists/Antagonists in Breast Cancer Therapy: A Critical Review. *Bioorganic Chem.* **2017**, *71*, 257–274. [CrossRef]
58. Thomas, M.P.; Potter, B.V.L. Estrogen O-Sulfamates and Their Analogues: Clinical Steroid Sulfatase Inhibitors with Broad Potential. *J. Steroid Biochem. Mol. Biol.* **2015**, *153*, 160–169. [CrossRef]
59. Pang, X.; Fu, W.; Wang, J.; Kang, D.; Xu, L.; Zhao, Y.; Liu, A.-L.; Du, G.-H. Identification of Estrogen Receptor α Antagonists from Natural Products via In Vitro and In Silico Approaches. *Oxid. Med. Cell. Longev.* **2018**, *2018*, 6040149. [CrossRef] [PubMed]
60. Deshpande, S.H.; Muhsinah, A.B.; Bagewadi, Z.K.; Ankad, G.M.; Mahnashi, M.H.; Yaraguppi, D.A.; Shaikh, I.A.; Khan, A.A.; Hegde, H.V.; Roy, S. In Silico Study on the Interactions, Molecular Docking, Dynamics and Simulation of Potential Compounds from *Withania somnifera* (L.) Dunal Root against Cancer by Targeting KAT6A. *Molecules* **2023**, *28*, 1117. [CrossRef]
61. Kostadinova, A.; Hazarosova, R.; Topouzova-Hristova, T.; Moyankova, D.; Yordanova, V.; Veleva, R.; Nikolova, B.; Momchilova, A.; Djilianov, D.; Staneva, G. Myconoside Interacts with the Plasma Membranes and the Actin Cytoskeleton and Provokes Cytotoxicity in Human Lung Adenocarcinoma A549 Cells. *J. Bioenerg. Biomembr.* **2022**, *54*, 31–43. [CrossRef]
62. Kostadinova, A.; Staneva, G.; Topouzova-Hristova, T.; Moyankova, D.; Yordanova, V.; Veleva, R.; Nikolova, B.; Momchilova, A.; Djilianov, D.; Hazarosova, R. Myconoside Affects the Viability of Polarized Epithelial MDCKII Cell Line by Interacting with the Plasma Membrane and the Apical Junctional Complexes. *Separations* **2022**, *9*, 239. [CrossRef]
63. Zheleva-Dimitrova, D.; Nedialkov, P.; Giresser, U. A Validated HPLC Method for Simultaneous Determination of Caffeoyl Phenylethanoid Glucosides and Flavone 8-C-Glycosides in *Haberlea rhodopensis*. *Nat. Prod. Commun.* **2016**, *11*, 791–792. [CrossRef] [PubMed]
64. Patel, K.; Patel, D.K. Medicinal Importance, Pharmacological Activities, and Analytical Aspects of Hispidulin: A Concise Report. *J. Tradit. Complement. Med.* **2016**, *7*, 360–366. [CrossRef] [PubMed]
65. Wang, Y.; Guo, S.; Jia, Y.; Yu, X.; Mou, R.; Li, X. Hispidulin Inhibits Proliferation, Migration, and Invasion by Promoting Autophagy via Regulation of PPARγ Activation in Prostate Cancer Cells and Xenograft Models. *Biosci. Biotechnol. Biochem.* **2021**, *85*, 786–797. [CrossRef]
66. Sun, Y.; Duan, X.; Wang, F.; Tan, H.; Hu, J.; Bai, W.; Wang, X.; Wang, B.; Hu, J. Inhibitory Effects of Flavonoids on Glucose Transporter 1 (GLUT1): From Library Screening to Biological Evaluation to Structure-Activity Relationship. *Toxicology* **2023**, *488*, 153475. [CrossRef] [PubMed]
67. Chang, C.-J.; Hung, Y.-L.; Chen, T.-C.; Li, H.-J.; Lo, Y.-H.; Wu, N.-L.; Chang, D.-C.; Hung, C.-F. Anti-Proliferative and Anti-Migratory Activities of Hispidulin on Human Melanoma A2058 Cells. *Biomolecules* **2021**, *11*, 1039. [CrossRef]
68. Kim, H.A.; Lee, J. Hispidulin Modulates Epithelial-Mesenchymal Transition in Breast Cancer Cells. *Oncol. Lett.* **2021**, *21*, 155. [CrossRef] [PubMed]

Disclaimer/Publisher's Note: The statements, opinions and data contained in all publications are solely those of the individual author(s) and contributor(s) and not of MDPI and/or the editor(s). MDPI and/or the editor(s) disclaim responsibility for any injury to people or property resulting from any ideas, methods, instructions or products referred to in the content.

Article

Rapid and Simultaneous Extraction of Bisabolol and Flavonoids from *Gymnosperma glutinosum* and Their Potential Use as Cosmetic Ingredients

Mayra Beatriz Gómez-Patiño [1], Juan Pablo Leyva Pérez [2], Marcia Marisol Alcibar Muñoz [3], Israel Arzate-Vázquez [1] and Daniel Arrieta-Baez [1,*]

[1] Instituto Politécnico Nacional, Centro de Nanociencias y Micro y Nanotecnologías, Unidad Profesional Adolfo López Mateos, Av. Luis Enrique Erro S/N, Colonia Zacatenco, Mexico City 07738, Mexico; mbgomez@ipn.mx (M.B.G.-P.); iarzate@ipn.mx (I.A.-V.)

[2] Instituto Politécnico Nacional, Escuela Superior de Ingeniería Química e Industrias Extractivas, Unidad Profesional Adolfo López Mateos, Av. Luis Enrique Erro S/N, Colonia Lindavista, Mexico City 07738, Mexico; leyva.john022@gmail.com

[3] Instituto Politécnico Nacional, Escuela Nacional de Ciencias Biológicas, Unidad Profesional Adolfo López Mateos, Av. Wilfrido Massieu 399, Colonia Industrial Vallejo, Mexico City 07738, Mexico; malcibar@ipn.mx

* Correspondence: darrieta@ipn.mx; Tel.: +52-1-55-5729-6000 (ext. 57507)

Abstract: *Gymnosperma glutinosum* is a plant popularly known as "popote", "tatalencho", "tezozotla" or "pegajosa", and it is used in traditional medicine in the region of Tehuacán, Puebla (Mexico), for the treatment of jiotes and acne and to cure diarrhea using the aerial parts in infusions. To analyze the phytochemical composition, we have developed a rapid protocol for the extraction and separation of the components of the aerial parts of *G. glutinosum*. After a maceration process, chloroformic and methanolic extracts were obtained and analyzed. Extracts were evaluated by GC-MS (gas chromatography-mass spectrometry), and their composition revealed the presence of (−)-α-bisabolol (BIS) as the main component in the chloroformic extract, which was isolated and analyzed by ^{1}H NMR to confirm its presence in the plant. The analysis of methanolic extracts by UPLC-MS (ultra-performance liquid chromatography-mass spectrometry) revealed the occurrence of six methoxylated flavones with m/z 405.08 ($C_{19}H_{18}O_{10}$), m/z 419.09 ($C_{20}H_{20}O_{10}$) and m/z 433.11 ($C_{21}H_{22}O_{10}$), and a group of C20-, C18-hydroxy-fatty acids, which give the plant its sticky characteristic. The presence of BIS, an important sesquiterpene with therapeutic skin effects, as well as some antioxidant compounds such as methoxylated flavones and their oils, could play an important role in cosmetology and dermatology formulations.

Keywords: *Gymnosperma glutinosum*; cosmetology; skin care; antioxidants; flavonoids; bisabolol

1. Introduction

The human skin is an organ that covers 15% of the total weight of the human body, which is not only important for aesthetic reasons, but also because it is responsible for many vital functions, among them the protection against external factors, regulation of fluid balance, metabolism, elimination of toxins and body shape maintenance [1,2]. Most people care about maintaining healthy skin, which promotes mental health by increasing people's self-confidence [3–5]. The use of cosmetics has become essential in our society, and although many plant products have been replaced by synthetic chemical compounds, a replacement has not been found for the safety and efficacy of natural products, which is why in recent years, the preference for natural products has resurfaced [6–10]. Since the awareness of the long-term benefits of natural ingredients in cosmetic products is increasing, they are being considered more, and recent studies indicate that plant components, such as phenolics, flavonoids and polysaccharides, have a high potential for cosmetic applications [11–15]. Besides the presence of compounds that demonstrated certain benefits, it is important to consider their synergy. Extracts from medicinal plants are rich in compounds that act synergistically, and it is important to study these compounds in their natural percentage to understand their optimal biological activity [15]. Thus, in mild skin disorders, the topical application of certain preparations based on natural products, such as infusions, creams, balms and tinctures, having in mind these concepts, can be effective in preventing the development of more severe diseases.

Ethnobotanical studies have helped us to understand the use of plants in traditional medicine [16]. These studies have not only helped us to understand the historical relevance of plants but also the importance they play in human health. Based on this type of study, it was determined that *G. glutinosum*, a plant that is popularly known as "popote", "tatalencho", "pegajosa" or "tezozotla", is used in traditional medicine in the region of Tehuacán, Puebla (Mexico) [17–21], for the treatment of jiotes and acne and in infusions to cure diarrhea. In recent years, ethnobotanical studies showed that *G. glutinousm* is one of the most important plants used in traditional medicine from the Tehuacan-Cuicatlan Biosphere Reserve, San Rafael, Coxcatlan, and Zapotitlan Salinas, Puebla (Mexico), for the treatment of diarrhea. Phytochemical studies of *G. glutinosum* have demonstrated the presence of essential oils, flavonoids and diterpenes. Most of the compounds isolated from *G. glutinosum* are methoxylated flavones such as 5,7-dihydroxy 3,6,8-trimethoxyflavone and 5,7-dihydroxy 3,6,8,2′,4′,5′-hexamethoxyflavone, and some ent-labdane-type diterpene [20–23]. More recent studies showed the isolation of a diterpene ent-labdene-type: ent-dihydrotumanoic acid (DTA) with antidiarrheal and antinociceptive effects [24–26]. However, even when this plant is used in the treatment of some skin problems, there are no studies about the relation between the isolated compounds and these diseases.

(−)-α-bisabolol (BIS), a sesquiterpene alcohol, has been mostly isolated from chamomile [27], and there are no reports of the presence in the genus *Gymnosperma*. BIS has different biological activities, including antioxidant, anticancer, anti-inflammatory, anti-infection, and skin-shoothing and -moisturizing properties [28–31]. At present, BIS is mainly manufactured through the steam-distillation of chamomile essential oils. Some products are produced by synthesized BIS; however, the process requires an additional economically unviable purification step due to the presence of the diastereomer (+)-α-bisabolol [32]. Therefore, finding new natural sources of bisabolol is essential to specialty pharmacological and cosmetologically industries.

In this work, based on ethnobotanical studies, we studied *G. glutinosum* to find compounds as ingredients for cosmetic purposes, and three groups of compounds were identified under the bases of GC-MS and UPLC-MS analyses. (−)-α-bisabolol and 6-epi-shyobunol, six methoxylated flavones, and a group of C20-, C18-hydroxy-fatty acids were identified in this extract and represent interesting mixtures for further investigations in the cosmetic industry.

2. Materials and Methods

2.1. Chemicals

Chloroform (CAS 110-54-3, 99%), acetonitrile (CAS 75-05-8, 99%), methanol (CAS 64-578-6, 99%), water (CAS 7732-18-5) and the ESI-TOF (electrospray ionization-time of flight) tuning mix calibrant were obtained from Supelco (Toluca, Edo. de México, México). The deuterated $CDCl_3$, MeOD and TMS were acquired from Sigma Aldrich (Toluca, Edo. de Mexico, Mexico).

2.2. UPLC and GC Coupled to Mass Spectrometry Analysis

Ultra-performance liquid chromatography-mass spectrometry (UPLC-MS) analysis was conducted in a Ultimate3000 UPLC system (Dionexcorp., Sunnyvale, CA, USA) with photodiode array detection (PAD), coupled to a Bruker MicrOTOF-QII system by an electrospray ionization (ESI) interface (Bruker Daltonics, Billerica, USA). Mobile phase used in the system consisted of 0.1% formic acid in water (A) and acetonitrile (B) using a gradient program of 5–35% (B) in 0–10 min, 35–80% (B) in 10–10.1 min, 80–80% (B) in

10.1–11, 80–45% (B) in 11–11.1, 45–5% (B) in 11.1–12 min and 5% (B) in 12–15 min. The chromatographic column used was a Hypersil C18 column (3.0 µm, 125 × 4.0 mm) (Varian). The solvent flow rate used was 0.5 mL/min, and the column temperature was set to 30 °C. For the mass spectrometer, conditions in the negative mode were as follows: drying gas (nitrogen), flow rate, 8 L/min; gas temperature, 180 °C; scan range, 50–3000 m/z; end plate off-set voltage, -500 V; capillary voltage, 4500 V; and nebulizer pressure, 2.5 bar.

Direct injection electrospray ionization-mass spectrometry (DI-ESI-MS) analyses were conducted on Bruker MicrOTOF-QII system by an electrospray ionization interface (Bruker Daltonics, Billerica, MA, USA) operating in the negative ion mode (ESI-).

A total of 1 mg of the sample was resuspended in 1 mL of methanol, filtered through a 0.25 µm polytetrafluoroethylene (PTFE), and diluted 1:100 with methanol to avoid saturation of the capillary and cone soiling. To inject the sample directly into the spectrometer, a 74900-00-05 Cole Palmer syringe pump (Billerica, MA, USA) was used and set at 8 µL/min to obtain a constant flow rate. The capillary voltage was set to 4500 V, and nitrogen was used as a drying and nebulizing gas, using a flow rate of 4 L/min (0.4 Bar) with a gas temperature of 180 °C. The spectrometer was calibrated with an ESI-TOF tuning mix calibrant (Sigma-Aldrich, Toluca, Estado de Mexico, Mexico).

To analyze compounds' structure, a tandem mass spectrometry (MS/MS) analysis was performed using negative electrospray ionization with the appropriate mass set. According to the obtained pattern, suitable fragments were analyzed by a Bruker Compass Data Analysis 4.0 (Bruker Daltonics), which provided a list of possible elemental formulas using Generate Molecular Formula Editor, as well as a sophisticated comparison of the theoretical with the measured isotope pattern (σ value) for increased confidence in the suggested molecular formula (Bruker Daltonics Technical Note 008, 2004). The accuracy threshold for confirmation of elemental compositions was established at 5 ppm.

Gas chromatography analysis was performed using a gas chromatograph 456-GC SCION TQ (Bruker, Biellerica MA, USA). The injector port was set at 220 °C with split 10. Separation was achieved using an RXI-5SIL (Fused silica, 30 m × 0.32 mm (ResteK) and helium at a flow rate of 1 mL/min. Column oven was programmed in the following conditions: 55 °C for 1.0 min, 55 °C to 155 °C at 20 °C/min, 155 °C for 2 min, 155 °C to 255 °C at 10 °C/min, 255 °C for 5.0 min, and finally from 255 °C to 280 °C at 10 °C/min and 280 °C for 5 min. The MS was set in TIC mode with an EI (electron ionization) of 70 eV.

2.3. NMR Spectroscopy

^{1}H nuclear magnetic resonance (^{1}H NMR) experiments of the soluble products were conducted on a Bruker Instruments ASCEND 750 spectrometer (Billerica, MA, USA). The resonance frequency was 750.12 MHz, with a typical acquisition time of 2.1845 s and a delay time of 1.0 s between successive acquisitions. The ^{1}H and ^{13}C chemical shifts are given in units of δ (ppm) relative to tetramethylsilane (TMS), where δ (TMS) belongs to 0 ppm.

3. Results

Aerial parts of *G. glutinosum* were collected in Santa Maria La Alta, a village in the municipality of Tehuacan, Puebla (Mexico) (18.60001:−97.65754) (Figure 1). A total of 800 g of the plant were collected in two different months to standardize the extraction process and determine if there could be some changes in the production of the phytochemical compounds. The first collection was completed in June and the second one in December.

Figure 1. Presence of *G. glutinosum* ("popote, pegajosa") in the Tehuacan Valley.

3.1. Isolation and Standardization of the Method of Extraction

With plants collected in June and December 2022, the next procedure was applied: 250 g of the aerial parts of the fresh plant material were ground and extracted two times with 500 mL of chloroform at room temperature. After filtration, the solvent was evaporated to obtain the chloroform extract. This procedure was completed twice for each plant material collected. The chloroform extract was partitioned with chloroform and methanol. Two partitioned fractions were obtained: a chloroform fraction 1 (CHCl$_3$ Fc1) and methanolic fraction 1 (MeOH Fc1). The yields are shown in Table 1.

Table 1. Yields of the fractions obtained from the *G. glutinosum* chloroform extraction.

Material Collected	Yield	CHCl$_3$ Fc1	MeOH Fc1
1 June 2022 (250 g)	18.9 g (7.56%) [1]	17.1 g	1.6 g
2 December 2022 (250 g)	19.5 g (7.80%) [1]	17.6 g	1.80 g

[1] Extraction yield is the average of two processes.

To analyze if the extraction with chloroform had been exhausted, the residue of the plant material was subject to another extraction with methanol and the extract was subject to the same partition as those applied to the chloroform extract. Two fractions were obtained: a chloroform fraction 2 (CHCl$_3$ Fc2) and a methanolic fraction 2 (MeOH Fc2). The yields are shown in Table 2.

Table 2. Yields of the fractions obtained from the *G. glutinosum* methanolic extraction.

Material Collected	Yield	CHCl$_3$ Fc2	MeOH Fc2
1 June 2022 (250 g)	28.3 g (11.32%) [1]	5.1 g	20.4 g
2 December 2022 (250 g)	27.5 g (11.00%) [1]	7.6 g	19.8 g

[1] Extraction yield is the average of two processes.

3.2. Chemical Composition Assay by GC-MS of the CHCl$_3$ Fc1

The fraction partitioned with chloroform was analyzed through GC-MS. Samples were analyzed by triplicate, injecting 1 mL, and carried out at 1 mL·min^{-1} by ultrapure helium. Two peaks were detected as the main compounds, and the mass spectra of each molecule were compared with those in the NIST (National Institute of Standards and Technology) database software. Under these conditions, 6-epi-shyobunol and (−)-α-bisabolol (BIS) were detected at 2.5 and 97.5%, respectively (Figure 2).

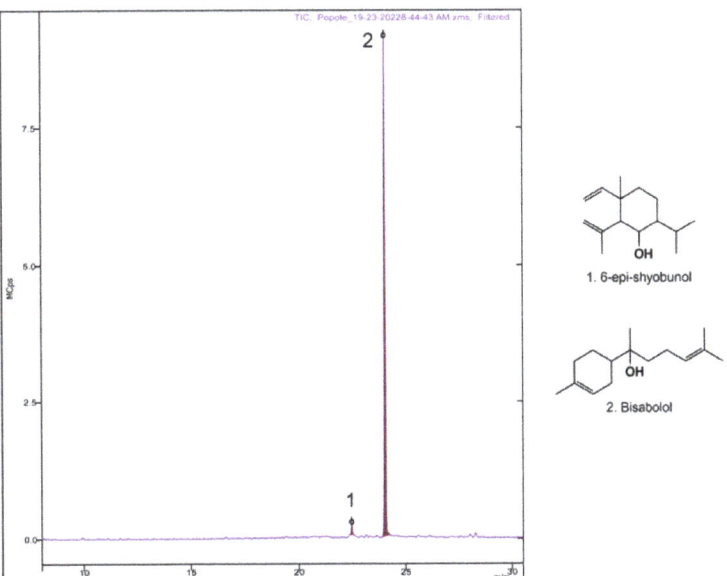

Figure 2. Chromatogram of the GC-MS analysis and compounds detected in the CHCl₃ Fc1 from *G. glutinosum*.

From the methanol extraction, chloroform partitioned extract (CHCl₃ Fc2) was analyzed under the same conditions. However, no compounds were detected indicating that BIS was completely extracted with chloroform.

3.3. Chemical Composition by UPLC-MS Assay of MeOH Fc1

MeOH Fc1 was subject to a UPLC-MS analysis and according to Figure 3, different peaks were detected and analyzed by the extracted ion chromatogram (EIC) from the total ion chromatograph (TIC) to generate two main groups of compounds detected in this extract. In the first group, two types of long-chain fatty acids were present for the m/z 321.23 and 319.21. For the detected molecular ion at m/z 321.23 [M-H]$^{-1}$, a molecular formula $C_{20}H_{34}O_3$ was assigned. According to Figure 3(1), at least six peaks correspond to this molecular weight indicating the presence of isomers. Under the bases of *ms/ms* analysis, only the main peak was identified as 20-hydroxyicosa-(5,8,11)-trienoic acid (Figure 3(1)). In the same regard, the molecular ion at m/z 319.21 [M-H]$^{-1}$ was consisting of a molecular formula $C_{18}H_{32}O_3$. In this case, four peaks were detected, indicating the presence of the same number of isomers that could be assigned to 20-hydroxyicosa-(5,8,11,14)-tetraenoic acid derivatives (Figure 3(2)).

For the second group, a methoxylated flavones group was detected and analyzed. Two peaks at Rt of 11.9 and 13.5 min (Figure 3(3)) showed the same molecular ion at m/z 405.08, which consisted of the molecular formula $C_{19}H_{18}O_{10}$ (5,7,2′,4′-Tetrahydroxy-3,6,8,5′-tetramethoxyflavone and 5,7,4′,5′-Tetrahydroxy-3,6,8,2′-tetramethoxyflavone). m/z 419.09 detected an Rt of 13.1 and 13.9 min (Figure 3(4)) and consisted of the molecular formula of $C_{20}H_{20}O_{10}$ for other two methoxylated flavones (5,7,2′-Trihydroxy-3,6,8,4′,5′-pentamethoxyflavone and 5,7,4′-Trihydroxy-3,6,8,2′,5′-pentamethoxyflavone). Finally, another two peaks at Rt of 14.3 and 15.4 min (Figure 3(5)) showed the same molecular ion at m/z 433.11, assigned to the molecular formula $C_{21}H_{22}O_{10}$ (5,7-Dihydroxy-3,6,8,3′,4′,5′-hexamethoxyflavone and 2-(5-Hydroxy-2,3-dimethoxyphenyl)-5-hydroxy-3,6,7,8-tetramethoxy-4H-1-benzopyran-4-one).

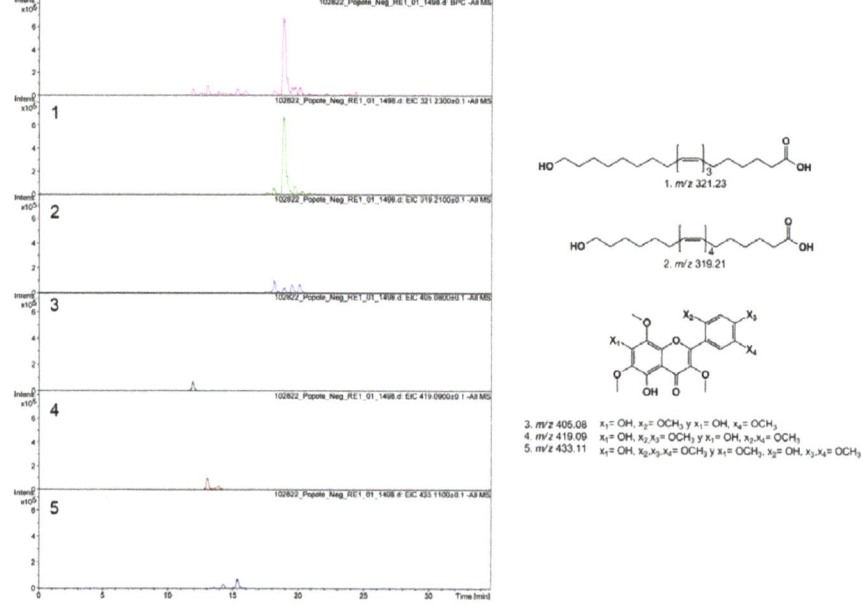

Figure 3. UPLC-MS chromatogram (upper chromatogram, TIC) and EIC-chromatogram analysis of the compounds detected in MeOH Fc1 from *G. glutinosum*. (1. EIC m/z 321, 2. EIC m/z 319, 3. m/z 405, 4. EIC m/z 419 and 5. EIC m/z 433).

The retention time (Rt), formula, name and relative percentage of the compounds detected are given in Table 3.

Table 3. Retention time and relative percentage of the compounds detected in the chromatogram of the UPLC-MS analysis of the MeOH Fc1 of *G. glutinosum*.

Rt	Formula	[M-H]$^-$ MW$_{detected}$	[M-H]$^-$ MW$_{exact}$	% Relative
11.9	$C_{19}H_{18}O_{10}$	405.0804	405.0816	7.0
13.1	$C_{20}H_{20}O_{10}$	419.0977	419.0972	11.0
13.6	$C_{20}H_{20}O_{10}$	405.0809	405.0816	1.1
13.9	$C_{20}H_{20}O_{10}$	419.0967	419.0972	4.1
14.3	$C_{21}H_{22}O_{10}$	433.1136	433.1129	3.0
15.4	$C_{21}H_{22}O_{10}$	433.1143	433.129	7.9
17.5	$C_{20}H_{32}O_3$	319.2197	319.2267	0.2
17.6	$C_{20}H_{34}O_3$	321.2393	321.2424	0.7
18.1	$C_{20}H_{32}O_3$	319.2195	319.2267	5.1
18.9	$C_{20}H_{34}O_3$	321.2399	321.2424	44.4
19.5	$C_{20}H_{32}O_3$	319.2191	319.2267	3.4
19.7	$C_{20}H_{34}O_3$	321.2394	321.2424	3.0
20.1	$C_{20}H_{32}O_3$	319.2197	319.2267	6.2
20.3	$C_{20}H_{34}O_3$	321.2395	321.2424	0.8
20.8	$C_{20}H_{34}O_3$	321.2398	321.2424	0.8

Under these conditions, BIS was not detected in the methanolic fraction.

The methanol extract, which was partitioned into a chloroform fraction ($CHCl_3$ Fc2) and a methanolic fraction (MeOH Fc2), was analyzed by means of DIESI-MS. As we can see in Figure 4, small peaks corresponding to the methoxylated flavones were detected, especially peaks with molecular ions at m/z 321, 405, and 419.

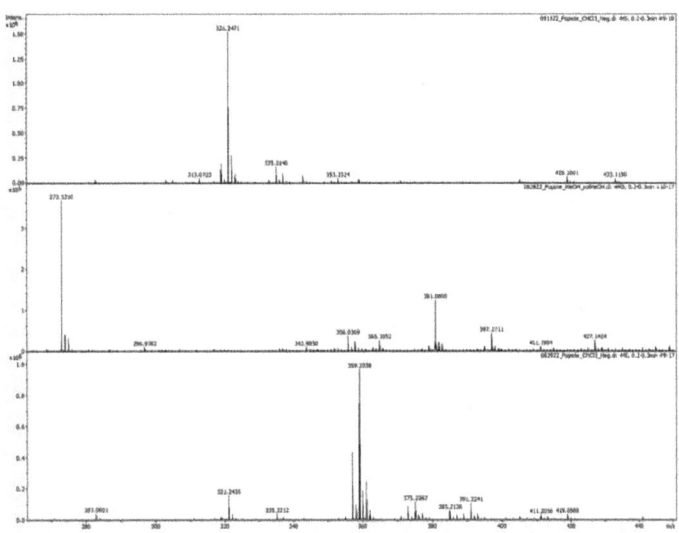

Figure 4. DIESI-MS analysis of the MeOH Fc2 from *G. glutinosum* compared with that obtained from MeOH Fc1. Upper spectrum: MeOH Fc1; middle spectrum: MeOH Fc2 and lower spectrum: CHCl$_3$ Fc2.

3.4. NMR Analysis of the Chloroform Extract of G. glutinosum

CHCl$_3$ Fc1 was analyzed through ^{1}H NMR. In Figure 5, a comparison of simulated and obtained spectra is shown. Vinylic protons at δ 5.2 and 5.4 ppm, as well as the methyl groups at δ 1.3 and 1.7 ppm confirm the presence of BIS, which was detected before by means of GC-MS.

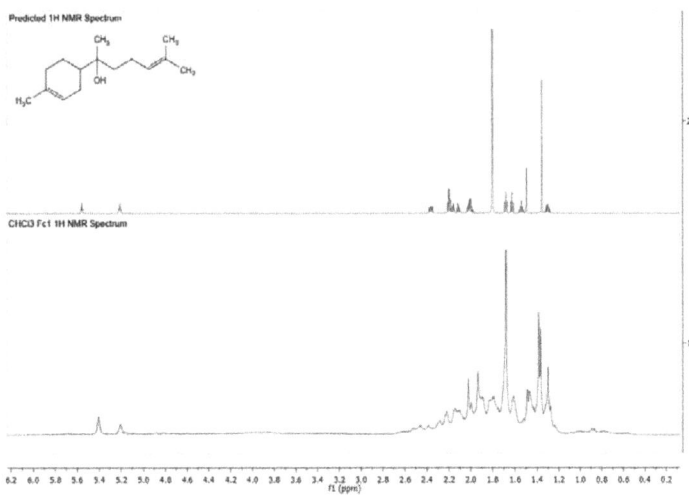

Figure 5. ^{1}H NMR spectra of CHCl$_3$ Fc1 from *G. glutinosum* (lower spectra), compared with a predicted spectrum of BIS (software MestReNova ver 6.0.2).

4. Discussion

In addition to the use of *G. glutinosum* in traditional medicine in the Tehuacan Valley for stomach problems, it has been used to help with some skin problems. Under these ethnobotanical studies, this plant that grows in the wild seems to have promising active compounds, which with adequate procedures implemented for their extraction could provide a high-added value raw material source for natural antioxidant constituents with a high potential for application in cosmetology and dermatology formulations. In recent years, different studies have been conducted to incorporate plants or herbal extracts for their therapeutic potential in the market as skincare products [14,15]. In the same way, phytochemical compounds have been evaluated in vivo and in vitro models to analyze their biological activity. However, it is necessary to consider the synergy of different extracts and evaluate the activity as a complex extract because it is possible that it could have a more effective impact.

To address this objective, two extractions were performed to obtain the most compounds and characterize them under different analytical techniques.

As a plant that grows in the wild, it was important to know if the phytochemical compounds are present at any time of the year, at least in this region of México. Procedures were applied in the same way under the same conditions, and according to Tables 1 and 2, the yield in the extraction seems to be very similar.

The dry extract of chloroform was partitioned with chloroform (CHCl$_3$ Fc1) and methanol (MeOH Fc1), and CHCl$_3$ Fc1 was analyzed by GC-MS. The results obtained and previously discussed in this work suggest the presence of BIS. According to Figure 1, two peaks were identified as 6-epi-shyobunol and (−)-α-bisabolol (BIS) at 2.5 and 97.5%, respectively. To confirm this result, BIS was purified and analyzed by ^{1}H NMR. Figure 5 shows the spectra obtained and compared with a simulation of BIS in the software MestReNova. According to both analyses, the presence of this compound in *G. glutinosum* is confirmed for the first time. With the proposed methodology, BIS was extracted quickly and selectively with a good yield. So, it could be possible that the biological activity shown by this plant in skin problems can be attributed to BIS [30,31].

(−)-α-bisabolol (BIS), a monocyclic sesquiterpene alcohol, was isolated and identified for the first time from chamomile (*Matricaria chamomilla*) [27], and to date is mainly obtained from this natural source; although, it has also been isolated from the essential oil of other medicinal plants [31]. It is a compound that has been considered safe due to its low toxicity, and its effects have been widely studied in different models, indicating its potential beneficial actions [31]. BIS has a variety of biological activities, including antioxidant, gastroprotective, anti-infection and anticancer properties. In atopic dermatitis, it has been found to help attenuated pruritus and inflamed skin. Other results include improved facial texture, skin oiliness, hydrated skin, brightness and better appearance in patients who used it as a treatment [28–31].

Even though the phytochemistry of *G. glutinosum* is poorly described, and there is no report of suitable compounds that could be used as cosmetic ingredients besides BIS, some methoxylated flavones were identified. From the partitioned fraction of methanol (MeOH Fc1), a methoxylated flavones group was detected and analyzed. Two peaks at Rt of 11.9 and 13.5 min (Figure 3(3)) showed the same molecular ion at m/z 405.08, which consisted of the molecular formula $C_{19}H_{18}O_{10}$ (5,7,2′,4′-Tetrahydroxy-3,6,8,5′-tetramethoxyflavone and 5,7,4′,5′-Tetrahydroxy-3,6,8,2′-tetramethoxyflavone). m/z 419.09 detected an Rt of 13.1 and 13.9 min (Figure 3(4)) and consisted of the molecular formula of $C_{20}H_{20}O_{10}$ for two other methoxylated flavones (5,7,2′-Trihydroxy-3,6,8,4′,5′-pentamethoxyflavone and 5,7,4′-Trihydroxy-3,6,8,2′,5′-pentamethoxyflavone). Finally, another two peaks at Rt of 14.3 and 15.4 min (Figure 3(5)) showed the same molecular ion at m/z 433.11, assigned to the molecular formula $C_{21}H_{22}O_{10}$ (5,7-Dihydroxy-3,6,8,3′,4′,5′-hexamethoxyflavone and 2-(5-Hydroxy-2,3-dimethoxyphenyl)-5-hydroxy-3,6,7,8-tetramethoxy-4H-1-benzopyran-4-one).

Most of these compounds have been previously reported in *G. glutinocum*. 5,7-dihydroxy-3,6,8,2′,4′,5-hexamethoxyflavone was isolated with (−)-17-hydroxy-neo-clerod-3-en-15-oic acid by Canales et al. from two samples of *G. glutinosum* obtained in Puebla and Hidalgo State (Mexico) [20,21]. Extracts obtained from these samples showed antimicrobial activity. In 2009, Serrano et al. [22] described the isolation of two methoxylated flavones, 5,7-dihydroxy-3,6,8-trimethoxyflavone and 5,7-dihydroxy-3,6,8,2′,4′,5′-hexamethoxyflavone, which were responsible for the fungal activity against *Aspergillus niger*, *Candida albicans*, *Fusarium sporotrichum* and *Trichophyton mentagrophytes*. Some other flavonoids, such as quercitrin, quercetin, kaempferol, rutin and vitexin have been described from *G. glutinosum* [23]. All these compounds have demonstrated their biological activity as antifungal, antimicrobial and antioxidant compounds and these activities contribute to the medicinal properties, which are used in traditional medicine in the Tehuacan, Puebla (Mexico) region. More recently, a diterpene ent-labdene-type: ent-dihydrotumanoic acid (DTA) with antidiarrheal and antinociceptive effects was isolated.

The increase in knowledge of the damage that ultraviolet radiation can cause in carcinogenesis and aging has increased the use of skin care products, especially sunscreens. However, most of these products are made with synthetic molecules or chemical substances that cause dermal toxicity. Nowadays, different reports of the beneficial effects of plants and herbal products for the skin are available [33]. Most of the plants are rich in polyphenols, flavonoids and some other compounds with antioxidant activity that can protect the skin from the effects of ultraviolet radiation. These herbal extracts should be considered for use in herbal skincare products.

Antioxidant molecules, such as methoxylated flavones, can be used to reduce and neutralize free radicals and when combined with recognized natural compounds like BIS, could improve their biological activity, resulting in their use as potential extracts implemented in cosmetic, pharmaceutical and therapeutic formulations [15,33].

One of the important characteristics of this plant is its sticky property, which is why it receives the same name in many places: "planta pegajosa (sticky plant)". This characteristic could be related to the presence of oils, waxes or fatty acids. According to the UPLC-MS analysis, two families of hydroxy-C18 and -C20 fatty acids were identified. For the detected molecular ion at m/z 321.23 $[M-H]^{-1}$, a molecular formula $C_{20}H_{34}O_3$ was assigned, and at least six peaks correspond to this molecular weight indicating the presence of isomers of the identified 20-hydroxyicosa-(5,8,11)-trienoic acid (Figure 3(1)). In the same regard, for the molecular ion at m/z 319.21 $[M-H]^{-1}$, four peaks were detected, indicating the presence of the same number of isomers that could be assigned to 20-hydroxyicosa-(5,8,11,14)-tetraenoic acid derivatives (Figure 3(2)).

The long-chain fatty acids present in the leaves of *G. glutinosum*, which are described for the first time in this work, could give the extract its waxy characteristic. Compounds such as waxes and oils extracted from plants have attracted attention for their properties to form films that can be used in cosmetic masks for skincare [34]. In this case, the presence of hydroxy long-chain fatty acids could help to keep the methoxylated flavones and the BIS in an aliphatic matrix for easier interaction with the skin. Anyway, functional groups such as hydroxyl and carbonyl groups in the long-chain fatty acids gave tunable properties to incorporate bioactive compounds for potentially sustainable alternatives over conventional products.

So, the presence of BIS, the antioxidant, the bacterial properties of the methoxylated flavones, and the aliphatic long-chain fatty acids are interesting mixtures for further investigation in the cosmetic industry of this extract obtained from *G. glutinosum*.

5. Conclusions

The use of cosmetics has become indispensable in our society. Although many plant products have been replaced by synthetic chemical compounds, the safety and efficacy of natural products have not been replaced, and in recent years, the preference for the natural has re-emerged. From an ethnobotanical study, a rapid and successful extraction of bio-compounds from *G. glutinosum* was conducted. From the chloroform extract, two partitions were obtained and analyzed. From the partition 1 (CHCl3 Fc1), the extraction and characterization of (−)-α-bisabolol as the main compound was conducted, with 97.5% of relative abundance. From partition 2 (MeOH Fc1), two groups of hydroxy-C18 and -C20 long-chain fatty acids are described for the first time in this plant. For the C20, the molecular ion at m/z 321.23 $[M-H]^{-1}$, at least six peaks indicated the presence of isomers of the identified 20-hydroxyicosa-(5,8,11)-trienoic acid (49.7%). In the same regard, for the C18, the molecular ion at m/z 319.21 $[M-H]^{-1}$, four peaks were assigned to 20-hydroxyicosa-(5,8,11,14)-tetraenoic acid derivatives (14.9%). On the other hand, under the basis of the UPLC-MS analysis, a set of methoxylated flavones are described. Two compounds with m/z 405.08, which consisted of the molecular formula $C_{19}H_{18}O_{10}$ (5,7,2′,4′-Tetrahydroxy-3,6,8,5′-tetramethoxyflavone and 5,7,4′,5′-Tetrahydroxy-3,6,8,2′-tetramethoxyflavone) (8.1%). Two compounds with m/z 419.09 consisted of the molecular formula of $C_{20}H_{20}O_{10}$ for two other methoxylated flavones (5,7,2′-Trihydroxy-3,6,8,4′,5′-pentamethoxyflavone and 5,7,4′-Trihydroxy-3,6,8,2′,5′-pentamethoxyflavone) (15.1%). Finally, there were another two compounds with m/z 433.11, assigned to the molecular formula $C_{21}H_{22}O_{10}$ (5,7-Dihydroxy-3,6,8,3′,4′,5′-hexamethoxyflavone and 2-(5-Hydroxy-2,3-dimethoxyphenyl)-5-hydroxy-3,6,7,8-tetramethoxy-4H-1-benzopyran-4-one) (10.9%).

For those with a rapid, economical and efficient process of extraction, this extract could be eligible for further studies as ingredients for cosmetic purposes since they present a set of biocompounds with useful photoprotective, moisturizing, and skin-lightening properties.

Author Contributions: M.B.G.-P. and D.A.-B. conceived and designed the main ideas of this paper, carried out the GC-MS, UPLC-MS and DIESI-MS experiments, analyzed the experimental results, and wrote the paper. J.P.L.P. collected the plant samples and classified the plant. J.P.L.P., M.M.A.M. and I.A.-V. carried out the compounds extraction experiments and helped to discuss the results. I.A.-V. reviewed and edited the last version of the manuscript. The authors read and approved the final manuscript. Investigation, D.A.-B., J.P.L.P., M.M.A.M. and M.B.G.-P.; project administration, D.A.-B. and M.B.G.-P.; supervision, D.A.-B. and M.B.G.-P. All authors have read and agreed to the published version of the manuscript.

Funding: This research was funded by the Post-graduated Investigation Office of the National Polytechnic Institute (SIP-IPN, grants No. 20221534, 20232111, 20221454 and 20231544).

Data Availability Statement: All data is contained within the article.

Conflicts of Interest: The authors declare no conflict of interest. The funders had no role in the design of the study; in the collection, analyses, or interpretation of data; in the writing of the manuscript; or in the decision to publish the results.

References

1. Bos, J.D.; Kapsenberg, M.L. The Skin Immune System Its Cellular Constituents and their Interactions. *Immunol. Today* **1986**, *7*, 235–240. [CrossRef]
2. Bos, J.D.; Kapsenberg, M.L. The Skin Immune System: Progress in Cutaneous Biology. *Immunol. Today* **1993**, *14*, 75–78. [CrossRef]
3. Erarslan, Z.B.; Ecevit-genç, G.; Kültür, Ş. Medicinal plants traditionally used to treat skin diseases in turkey—Eczema, psoriasis, vitiligo. *J. Fac. Pharm. Ank. Univ.* **2020**, *44*, 137–166.
4. Bodeker, G.; Ryan, T.J.; Volk, A.; Harris, J.; Burford, G. Integrative skin care: Dermatology and traditional and complementary medicine. *J. Altern. Complement. Med.* **2017**, *23*, 479–486. [CrossRef]
5. Weiss, R.F.; Fintelmann, V. *Herbal Medicine*; Thieme Medicinal Publishers: Stuttgart, Germany; New York, NY, USA, 2000; pp. 293–314.
6. Gurib-Fakim, A. Medicinal Plants: Traditions of Yesterday and Drugs of Tomorrow. *Mol. Asp. Med.* **2006**, *27*, 1–93. [CrossRef]
7. Verma, S.; Singh, S.P. Current and future status of herbal medicines. *Veter-World* **2008**, *1*, 347–350. [CrossRef]
8. Shedoeva, A.; Leavesley, D.; Upton, Z.; Fan, C. Wound Healing and the Use of Medicinal Plants. *Evid. Based Complement. Altern. Med.* **2019**, *2019*, 2684108. [CrossRef]
9. Dawid-Pać, R. Medicinal plants used in treatment of inflammatory skin diseases. *Adv. Dermatol. Allergol.* **2013**, *3*, 170–177. [CrossRef]
10. Tabassum, N.; Hamdani, M. Plants used to treat skin diseases. *Pharmacogn. Rev.* **2014**, *8*, 52. [CrossRef]
11. Aburjai, T.; Natsheh, F.M. Plants used in cosmetics. *Phytother. Res.* **2003**, *17*, 987–1000. [CrossRef]
12. González-Minero, F.J.; Bravo-Díaz, L. The Use of Plants in Skin-Care Products, Cosmetics and Fragrances: Past and Present. *Cosmetics* **2018**, *5*, 50. [CrossRef]
13. Rocha-Filho, P.A.; Ferrari, M.; Maruno, M.; Souza, O.; Gumiero, V. In Vitro and In Vivo Evaluation of Nanoemulsion Containing Vegetable Extracts. *Cosmetics* **2017**, *4*, 32. [CrossRef]

14. Ajjoun, M.; Kharchoufa, L.; Merrouni, I.A.; Elachouri, M. Moroccan medicinal plants traditionally used for the treatment of skin diseases: From ethnobotany to clinical trials. *J. Ethnopharmacol.* **2022**, *297*, 115532. [CrossRef]
15. Selwyn, A.; Govindaraj, S. Study of plant-based cosmeceuticals and skincare. *S. Afr. J. Bot.* **2023**, *158*, 429–442. [CrossRef]
16. Cotton, C.M. *Ethnobotany: Principles and Applications*; John Wiley and Sons: London, UK, 1996; p. 434.
17. Argueta, V.A.; Cano, A.J. *Atlas de las Plantas de la Medicina Tradicional Mexicana*; Instituto Nacional Indigenista: Mexico City, México, 1994; pp. 1318–1319.
18. Arias, T.A.A.; Valverde, V.M.T.; Reyes, S.J. *Las plantas de la región de Zapotitlán Salinas, Puebla*; Instituto Nacional de Ecología: Mexico City, México, 2000; pp. 8–9, 22.
19. Hernández, T.; Canales, M.; Avila, J.G.; Durán, A.; Caballero, J.; Romo de Vivar, A.; Lira, R. Ethnobotany and antibacterial activity of some plants used in traditional medicine of Zapotitlán de las Salinas, Puebla (México). *J. Ethnopharmacol.* **2003**, *88*, 181–188. [CrossRef]
20. Canales, M.; Hernández, T.; Caballero, J.; Romo de Vivar, A.; Avila, G.; Duran, A.; Lira, R. Informat consensus factor and antibacterial activity of the medicinal plants used by the people of San Rafael Coxcatlán, Puebla, México. *J. Ethnopharmacol.* **2005**, *97*, 429–439. [CrossRef]
21. Canales, M.; Hernández, T.; Serrano, R.; Hernández, L.B.; Duran, A.; Ríos, V.; Sigrist, S.; Hernández, H.L.H.; Garcia, A.M.; Angeles-López, O.; et al. Antimicrobial and general toxicity activities of *Gymnosperma glutinosum*: A comparative study. *J. Ethnopharmacol.* **2007**, *110*, 343–347. [CrossRef]
22. Serrano, R.; Hernández, T.; Canales, M.; García-Bores, A.M.; Romo De Vivar, A.; Céspedes, C.L.; Avila, J.G. Ent-labdane type diterpene with antifungal activity from *Gymnosperma glutinosum* (Spreng.) Less. (Asteraceae). *Boletín Latinoam. Y Del Caribe De Plantas Med. Y Aromáticas* **2009**, *8*, 412–418.
23. Morado-Castillo, R.; Quintanilla-Licea, R.; Gomez-Flores, R.; Blaschek, W. Total Phenolic and Flavonoid Contents and Flavonoid Composition of Flowers and Leaves from the Mexican Medicinal Plant *Gymnosperma glutinosum* (Spreng.) Less. *Eur. J. Med. Plants* **2016**, *15*, 1–8. [CrossRef]
24. Alonso-Castro, A.J.; González-Chávez, M.M.; Zapata-Morales, J.R.; Verdinez-Portales, A.K.; Sánchez-Recillas, A.; Ortiz-Andrade, R.; Isiordia-Espinoza, M.; Martínez-Gutiérrez, F.; Ramírez-Morales, M.A.; Domínguez, F.; et al. Antinociceptive Activity of Ent-Dihydrotucumanoic Acid Isolated from Gymnosperma glutinosum Spreng Less. *Drug Dev. Res.* **2017**, *78*, 340–348. [CrossRef]
25. Alonso-Castro, A.J.; Arana-Argáez, V.E.; Deveze-Alvarez, M.A.; Chan-Zapata, I.; Torres-Romero, J.C.; Carranza-Álvarez, C.; Luna-Rubio, S.; González-Chávez, M.M.; Zapata-Morales, J.R.; Aragón-Martínez, O.H.; et al. Anti-inflammatory and diuretic effects of the diterpene ent-dihydrotucumanoic acid. *Drug Dev. Res.* **2019**, *80*, 800–806. [CrossRef]
26. González-Chávez, M.M.; Arana-Argáez, V.; Zapata-Morales, J.R.; Ávila-Venegas, A.K.; Alonso-Castro, A.J.; Isiordia-Espinoza, M.; Martínez, R. Pharmacological evaluation of 2-angeloyl ent-dihydrotucumanoic acid. *Pharm. Biol.* **2017**, *55*, 873–879. [CrossRef]
27. McKay, D.L.; Blumberg, J.B. A review of the bioactivity and potential health benefits of chamomile tea (*Matricaria recutita* L.). *Phytother. Res.* **2006**, *20*, 519–530. [CrossRef] [PubMed]
28. Murata, Y.; Kokuryo, T.; Yokoyama, Y.; Yamaguchi, J.; Miwa, T.; Shibuya, M.; Yamamoto, Y.; Nagino, M. The Anticancer Effects of Novel alpha-Bisabolol Derivatives Against Pancreatic Cancer. *Anticancer. Res.* **2017**, *37*, 589–598. [CrossRef]
29. Ortiz, M.I.; Cariño-Cortés, R.; Ponce-Monter, H.A.; Castañeda-Hernández, G.; Chávez-Piña, A.E. Pharmacological interaction of α-bisabolol and diclofenac on nociception, inflammation, and gastric integrity in rats. *Drug Dev. Res.* **2018**, *79*, 29–37. [CrossRef] [PubMed]
30. Cavalcante, H.A.O.; Silva-Filho, S.E.; Wiirzler, L.A.M.; Cardia, G.F.E.; Uchida, N.S.; Silva-Comar, F.M.S.; Bersani-Amado, C.A.; Cuman, R.K.N. Effect of (−)-alpha-Bisabolol on the Inflammatory Response in Systemic Infection Experimental Model in C57BL/6 Mice. *Inflammation* **2020**, *43*, 193–203. [CrossRef] [PubMed]
31. Eddin, L.B.; Jha, N.K.; Goyal, S.N.; Agrawal, Y.O.; Subramanya, S.B.; Bastaki, S.M.A.; Ojha, S. Health Benefits, Pharmacological Effects, Molecular Mechanisms, and Therapeutic Potential of α-Bisabolol. *Nutrients* **2022**, *14*, 1370. [CrossRef]
32. Han, G.H.; Kim, S.K.; Yoon, P.K.S. Fermentative production and direct extraction of (−)-α-bisabolol in metabolically engineered *Escherichia coli*. *Microb. Cell Fact.* **2016**, *15*, 185. [CrossRef]
33. Sharma, R.R.; Deep, A.; Abdullah, S.T. Herbal products as skincare therapeutic agents against ultraviolet radiation-induced skin disorders. *J. Ayurveda Integr. Med.* **2022**, *13*, 100500. [CrossRef]
34. Gaspar, A.L.; Gaspar, A.B.; Contini, L.R.; Silva, M.F.; Chagas, E.G.; Bahú, J.O.; Concha, V.O.; Carvalho, R.A.; Severino, P.; Souto, E.B.; et al. Lemongrass (Cymbopogon citratus)-incorporated chitosan bioactive films for potential skincare applications. *Int. J. Pharm.* **2022**, *628*, 122301. [CrossRef] [PubMed]

Disclaimer/Publisher's Note: The statements, opinions and data contained in all publications are solely those of the individual author(s) and contributor(s) and not of MDPI and/or the editor(s). MDPI and/or the editor(s) disclaim responsibility for any injury to people or property resulting from any ideas, methods, instructions or products referred to in the content.

MDPI
St. Alban-Anlage 66
4052 Basel
Switzerland
www.mdpi.com

Separations Editorial Office
E-mail: separations@mdpi.com
www.mdpi.com/journal/separations

Disclaimer/Publisher's Note: The statements, opinions and data contained in all publications are solely those of the individual author(s) and contributor(s) and not of MDPI and/or the editor(s). MDPI and/or the editor(s) disclaim responsibility for any injury to people or property resulting from any ideas, methods, instructions or products referred to in the content.

www.ingramcontent.com/pod-product-compliance
Lightning Source LLC
LaVergne TN
LVHW070611100526
838202LV00012B/622